Charles Davies, A. M. Legendre

Selected Propositions in Geometrical Constructions and

Applications of Algebra to Geometry

Charles Davies, A. M. Legendre

Selected Propositions in Geometrical Constructions and Applications of Algebra to Geometry

ISBN/EAN: 9783337155858

Printed in Europe, USA, Canada, Australia, Japan

Cover: Foto ©berggeist007 / pixelio.de

More available books at **www.hansebooks.com**

SELECTED PROPOSITIONS

IN

GEOMETRICAL CONSTRUCTIONS

AND

APPLICATIONS OF ALGEBRA TO GEOMETRY.

BEING

A KEY

TO THE APPENDIX OF

DAVIES' LEGENDRE.

A. S. BARNES & COMPANY,

NEW YORK, CHICAGO AND NEW ORLEANS.

1875.

DAVIES' COURSE OF MATHEMATICS.

IN THREE PARTS.

I. COMMON SCHOOL COURSE.

DAVIES' PRIMARY ARITHMETIC.—The fundamental principles displayed in Object Lessons.

DAVIES' INTELLECTUAL ARITHMETIC.—Referring all operations to the unit 1 as the only tangible basis for logical development.

DAVIES' ELEMENTS OF WRITTEN ARITHMETIC.—A practical introduction to the whole subject. Theory subordinated to Practice.

DAVIES' PRACTICAL ARITHMETIC.—A combination of Theory and Practice, clear, exact, brief, and comprehensive.

II. ACADEMIC COURSE.

DAVIES' UNIVERSITY ARITHMETIC.—Treating the subject exhaustively as a *science*, in a logical series of connected propositions.

DAVIES' ELEMENTARY ALGEBRA.—A connecting link, conducting the pupil easily from arithmetical processes to abstract analysis.

DAVIES' UNIVERSITY ALGEBRA.—For institutions desiring a more complete but not the fullest course in pure Algebra.

DAVIES' PRACTICAL MATHEMATICS.—The science practically applied to the useful arts, as Drawing, Architecture, Surveying, Mechanics, etc.

DAVIES' ELEMENTARY GEOMETRY.—The important principles in simple form, but with all the exactness of rigorous reasoning.

DAVIES' ELEMENTS OF SURVEYING.—Re-written in 1870. A simple and full presentation for Instruction and Practice.

III. COLLEGIATE COURSE.

DAVIES' BOURDON'S ALGEBRA.—Embracing Sturm's Theorem, and a most exhaustive course. Re-written, in 1873.

DAVIES' UNIVERSITY ALGEBRA.—A shorter course than Bourdon, for Institutions having less time to give the subject.

DAVIES' LEGENDRE'S GEOMETRY.—A standard work in this country and in Europe.

DAVIES' ANALYTICAL GEOMETRY.—A full course of Analysis, embracing the applications to surfaces of the second order.

DAVIES' DIFFERENTIAL AND INTEGRAL CALCULUS, on the basis of Continuous Quantity and Consecutive Differences.

DAVIES' ANALYTICAL GEOMETRY AND CALCULUS.—The shorter treatises, combined in one volume.

DAVIES' DESCRIPTIVE GEOMETRY.—With application to Spherical Trigonometry, Spherical Projections, and Warped Surfaces.

DAVIES' SHADES, SHADOWS, AND PERSPECTIVE.—A succinct exposition of the mathematical principles involved.

DAVIES' NATURE AND UTILITY OF MATHEMATICS, Logically considered.

DAVIES AND PECK'S MATHEMATICAL DICTIONARY, or Cyclopedia of Mathematics.

Entered according to Act of Congress, in the year 1875, by
CHARLES DAVIES,
In the Office of the Librarian of Congress, at Washington.

PREFACE.

THE applications of Mathematical Science, to the Mechanic Arts, have received special attention within the past few years; and no system of instruction, in any department, is now complete, unless it extends beyond the theoretical and into the field of Practical Knowledge.

Geometry, in its essence and structure, is more purely abstract than any other branch of science. In its logical structure, it is conversant only about space; and yet, the relations and principles which it develops, afford the only basis of the true Practical.

To extend Geometry to some of its most interesting and useful applications, an Appendix has been prepared and added to Legendre, embracing many Problems of Geometrical construction, and many applications of Algebra to Geometry: the whole being designed to explain and illustrate the methods of making Geometry a practical as well as a theoretical science.

It would be unjust to those giving instruction, to add to their daily labors, the additional one, of finding appropriate solutions to so many difficult problems: hence, a Key has been made for their special use, in which the best methods of construction and solution are fully given.

It is confidently hoped that this addition to a work which has been an accepted Text Book, both in this country and in Europe, for many years, may add something to its great value; and it would be pleasant to indulge the hope that it will be received with a portion of that great favor which has been extended to the original work.

FISHKILL-ON-HUDSON, }
    March, 1875. }

KEY.

PROPOSITION I.—*Show that the bisectrices of two adjacent angles are perpendicular to each other.*

DEMONSTRATION.—Let DCB and DCA be two adjacent angles, and CP and CQ be their bisectrices.

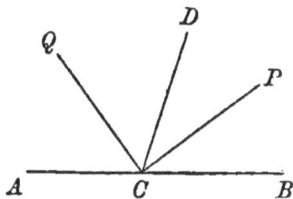

The angle PCD is equal to $\frac{1}{2}$ DCB by definition, and DCQ is equal to $\frac{1}{2}$ DCA: hence, PCD + DCQ = $\frac{1}{2}$ (DCB + DCA). The first member of this equation is equal to the angle QCP, and the second member is equal to a right angle (Bk. I. Prop. 1); hence, the angle QCP is a right angle, *which was to be proved.*

PROP. II.—*Show that the perimeter of any triangle is greater than the sum of the distances from any point within the triangle to its three vertices, and less than twice that sum.*

DEMONSTRATION.—Let ABC be any triangle, P any point within it, and PA, PB, PC, lines from P to the vertices.

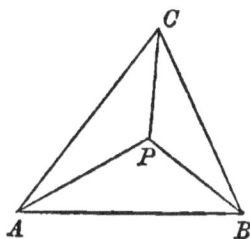

In accordance with Bk. I. Prop. 8, we have the following inequalities :

$$AB + BC > PA + PC$$
$$BC + CA > PA + PB$$
$$AB + AC > PB + PC.$$

Adding the first and second members of these inequalities, separately, and dividing by 2, we have,

$$AB + BC + CA > PA + PB + PC,$$

which proves the first part of the proposition.

In accordance with Bk. I, Prop. 7, we have the following inequalities :

$$AB < PA + PB$$
$$BC < PB + PC$$
$$CA < PC + PA.$$

Adding the members of these inequalities, separately, and then factoring, we have,

$$AB + BC + CA < 2 (PA + PB + PC),$$

which proves the second part of the proposition.

PROP. III.—*Show that the angle between the bisectrices of two consecutive angles of any quadrilateral is equal to one-half the sum of the other two angles.*

DEMONSTRATION.—Let ABDE be any quadrilateral, and let AC and BC be the bisectrices of the angles A and B.

The sum of the 4 angles of the quadrilateral is equal to 4 right angles (Bk. I. Prop. 26, Cor. 1) ; or, denoting the angles by the letters A, B, D, E, and 1 right angle by R, we have,

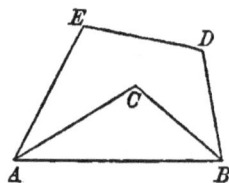

$$A + B + D + E = 4R.$$

Dividing both members of this equation by 2, factoring and transposing, we have,

$$\tfrac{1}{2} (D + E) = 2R - \tfrac{1}{2} (A + B) \quad . \quad . \quad . \quad (1).$$

But CAB and CBA are, respectively, equal to $\frac{1}{2}$ A and $\frac{1}{2}$ B; hence, (Bk. I. Prop. 25, Cor. 1) we have for the angle ACB, denoted by C,

$$C = 2R - \tfrac{1}{2}(A + B) \quad \cdot \quad \cdot \quad \cdot \quad (2).$$

Hence, in accordance with axiom 1, we have, from (1) and (2),

$$C = \tfrac{1}{2}(E + D),$$

which was to be proved.

Prop. IV. Theorem.—*Show that any point in the bisectrix of an angle is equally distant from the sides of the angle.*

Demonstration.—Let ABC be any angle, BP its bisectrix, and P any point of BP.

Draw PR and PQ respectively perpendicular to the sides BA and BC; then will PR and PQ be the distances from P, to BA, and BC. In the triangles BPR and BPQ, the angles PBR and PBQ are equal, by hypothesis, and the angles PRB and PQB are equal, because they are both right angles; hence, the angles BPR and BPQ are equal (Bk. I. Prop. 25, Cor. 2). The triangles BPR and BPQ, have the angles PBR and BPR respectively equal to the angles PBQ and BPQ, and the side BP common; hence, they are equal in all their parts (Bk. I. Prop. 6), that is, PR = PQ, *which was to be proved.*

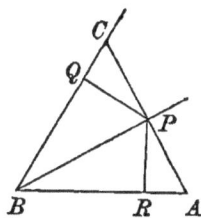

Conversely; If PQ=PR, the line PB is the bisectrix of the angle ABC.

Corollary.—If a line, as APC is drawn perpendicular to the bisectrix of an angle, the parts intercepted between it and the sides of the angle will be equal: for the triangles PQC and PRA are equal in all their parts , hence, PC equals PA.

PROP. V.—*If two sides of a triangle are prolonged beyond the third side, show that the bisectrices of the included angle and of the exterior angles, all meet in the same point.*

DEMONSTRATION.—Let ABC be any triangle, and let CO and BO be the bisectrices of the exterior angles ECB and DBC.

Draw AO, and also, draw OE perpendicular to AE, OD perpendicular to AD, and OQ perpendicular to BC. It is to be shown that AO is the bisectrix of the angle EAD.

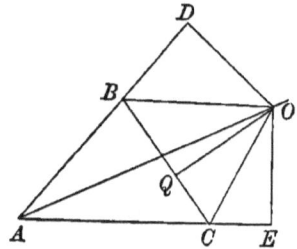

Because CO is the bisectrix of ECB, we have, OE = OQ, and because BO is the bisectrix of DBC, OD = OQ; consequently, (Ax. 1) OD = OE. Hence, from the last proposition, AO is the bisectrix of EAD, *which was to be proved.*

PROP. VI.—*Show that the projection of a line on a parallel line is equal to the line itself; and that the projection of a line on a line to which it is oblique, is less than the line itself.*

DEMONSTRATION.—Let PQ be the projection of the parallel line AB, and also the projection of the oblique line AC.

Because AB is parallel to PQ, and because AP and BQ are both perpendicular to PQ and consequently parallel, the figure ABPQ is a parallelogram; hence, AB = PQ, (Bk. I. Prop. 30), *which proves the first part of the proposition.*

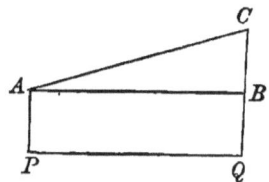

Because AB is perpendicular to QC, and AC is oblique to it, AC is greater than AB, (Bk. I. Prop. 15); hence AC is greater than PQ, *which proves the second part of the proposition.*

PROP. VII.—*If a line is drawn through the point of intersection of the diagonals of a parallelogram, show that the line is bisected at the point.*

DEMONSTRATION.—Let ABCD be a parallelogram, AC and BD its diagonals, and O their point of inter-section.

Draw the line PQ, through O, and limited by BC and AD.

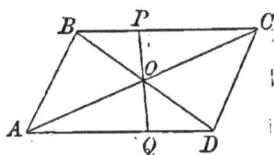

In the triangles POC and QOA, the sides AO and OC are equal (Bk. I. Prop. 31), the angles OCP and OAQ are equal (Bk. I. Prop. 20, Cor. 2), and the angles POC and QOA are equal, because they are opposite or vertical; hence, the triangles are equal in all their parts (Bk. I. Prop. 6): consequently, OP = OQ, *which was to be proved.*

PROP. VIII.—*The bisectrices of the four angles of any parallelogram form, by their intersection, a rectangle whose diagonals are parallel to the sides of the given parallelogram.*

DEMONSTRATION.—Let ABCD be any parallelogram, and let AS, BS, CR, and DR be the bisectrices of its angles.

Draw the diagonals PQ, RS, and prolong DA and BC to meet the bisectrices CR and AS, in L and T.

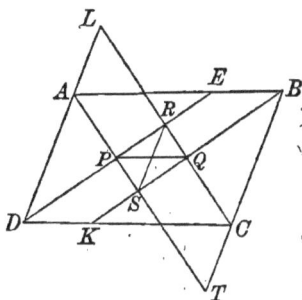

Because AS and BS are the bisectrices of the angles A and B, their included angle is equal to one-half the sum of the angles D and C (Prop. 3, Key); but the sum of D and C is two right angles; hence, the angle PSQ is a right angle. In like manner it may be shown that each of the angles SQR, QRP and RPS is a right angle; hence, PRQS is a rectangle.

Because DE and BK are parallels included between parallels, they are equal to each other (Bk. I. Prop. 28, Cor. 1). Since AP bisects the angle DAE and is perpendicular to DE, it divides the latter line into two equal parts, that is, EP is one-half of ED; in like manner it may be shown that BQ is one-half of BK; consequently EP is equal to BQ; hence, PQ is parallel to EB (Bk. I. Prop. 30). In the same way it may be shown that RS is parallel to LD, *which was to be proved.*

PROP. IX. THEOREM.—*Show that the sum of the distances from any point in the base of an isosceles triangle to the other two sides, is equal to the distance from the vertex of either angle at the base, to the opposite side.*

DEMONSTRATION.—Let ABC be an isosceles triangle and let P be any point in its base.

Draw CS parallel to AB, CD and QP perpendicular to AB, PR perpendicular to AC, and prolong QP to S.

The angle PCS is equal to ABC, because they are alternate angles, and the angle PCR is equal to ABC by hypothesis; hence, the angles PCS and PCR are equal. In the triangles PCS and PCR, the angles PCS and PCR are equal, as just shown, the angles PRC and PSC are equal, because both are right angles, and consequently the remaining angles CPR and CPS are equal: these triangles also have the side CP common; hence, they are equal in all their parts (Bk. I. Prop. 6); PR is therefore equal to PS, and QP + PR is equal to QP + PS, or

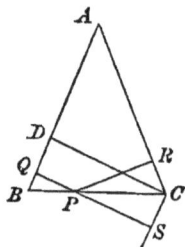

to QS; but QS is equal to DC (Bk. I. Prop. 28, Cor. 1); hence, PQ + PR = CD, *which was to be proved.*

PROP. X.—*Show that the middle point of the hypothenuse of a right-angled triangle is equally distant from the three vertices.*

DEMONSTRATION.—Let ABC be a right-angled triangle, and let O be the middle point of its hypothenuse.

Draw OB; also draw OP parallel to AB, and OQ parallel to CB.

In the triangles OQA and CPO, the angles COP and OAQ are equal (Bk. I. Prop. 19, Sch.); the angles OCP and AOQ are equal for the same reason, and the sides CO and OA are equal by hypothesis; hence, OP is equal to QA; but OP is also equal to BQ, and consequently BQ is equal to QA.—In the triangles OQA and OBQ, the side OQ is common, BQ is equal to QA, and the included angles OQB and OQA are equal because both are right angles; hence, OB is equal to OA, and to its equal to OC, *which was to be proved.*

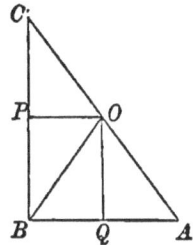

PROP. XI.—*Draw two lines that shall divide a given right angle into three equal parts.*

SOLUTION.—Let ADK be the given right angle. Take any distance DL, and on it as a side, construct an equilateral triangle DLQ, and bisect the angle QDL by the line DS, (Bk. III. Probs. 10 and 5). The angle QDL is equal to two-thirds of a right angle (Bk. I. Prop. 25, Cor. 5); hence, KDS and its equal SDQ, are each equal to one-

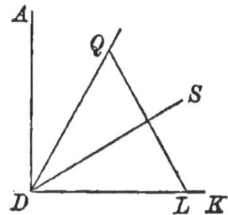

third of a right angle and the remaining angle QDA is also equal to one-third of a right angle. The required lines are, therefore, DS and DQ.

Prop. XII.—*Draw a line AP through the vertex A of a triangle ABF and perpendicular to the bisectrix of the angle A ; construct a triangle PBF having its vertex P on AP and its base coinciding with that of the given triangle : then show that the perimeter of PBF is greater than that of ABF.*

Demonstration.—Let ABF be the given triangle, AG the bisectrix of the angle A, AP perpendicular to AG, and let P be any point of AP. Prolong BA and make AM equal to AF; draw MP.

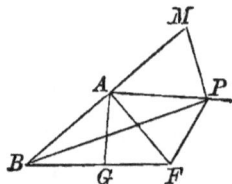

Because the angles BAF and FAM are adjacent, and AG is the bisectrix of the former, AP must be the bisectrix of the latter (Prop. 1, Key); hence the angles FAP and PAM are equal. In the triangles FAP and PAM, the side AP is common, AM is equal to AF by construction, and the included angles FAP and PAM are equal; hence, PM is equal to PF.—In accordance with Bk. I. Prop. 7, we have the inequality :

$$BP + PM > BA + AM.$$

Substituting PF for its equal PM, AF for its equal AM, and adding FB to each member, we have,

$$BP + PF + FB > BA + AF + FB,$$

which was to be proved.

Prop. XIII.—*Let an altitude of the triangle ABC be drawn from the vertex A, and also the bisectrix of the angle A ; then show that their included angle is equal to half the difference of the angles B and C.*

DEMONSTRATION.—Let ABC be a triangle, AQ the bisectrix of the angle A, and AP the altitude drawn from the vertex A. The angle CAB is equal to two right angles diminished by the sum of the angles B and C; hence, if we denote a right angle by R, we have,

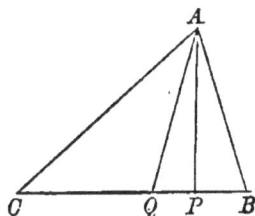

$$QAB \doteq R - \tfrac{1}{2}B - \tfrac{1}{2} C \quad . \quad . \quad . \quad (1).$$

In the right-angled triangle PAB, we have,

$$PAB = R - B \quad . \quad . \quad . \quad (2).$$

Subtracting (2) from (1), member from member, we have,

$$QAB - PAB = \tfrac{1}{2}B - \tfrac{1}{2} C, \text{ or } QAP = \tfrac{1}{2} (B - C),$$

which was to be proved.

PROP. XIV.—*Given two lines that would meet if sufficiently prolonged, to draw the bisectrix of their included angle without finding its vertex.* ●

SOLUTION.—Let AB and CD be the given lines.

Through any point P, of AB, draw PR parallel to CD (Bk. III. Prob. 6). Bisect the angle BPR by the line PQ, and draw the line PR perpendicular to PQ; bisect PR in S (Bk. III. Prob. 1), and draw SV parallel to PQ.

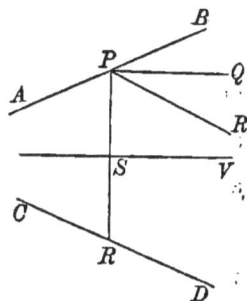

Because CD is parallel to PR, the angle between the given lines is equal to the angle RPB, and consequently the required bisectrix is parallel to PQ; hence, PR is perpendicular to that bisectrix. Since S is the middle point of PR, and SV is perpendicular to PR, it must be the required bisectrix (Prop. IV, Key, Cor.).

Prop. XV.—*From two points on the same side of a given line, draw two lines that shall meet each other at some point of the given line and make equal angles with that line.*

Solution.—Let P and Q be the given points and AK the given line.

Draw PN perpendicular to AK, and prolong it to M, making NM = NP; draw MQ, and from the point R in which it intersects AK, draw RP.

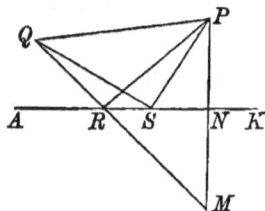

The right-angled triangles RNP and RNM, have the side PN equal to NM, by construction; the side RN common and the included angles PNR and MNR are equal; hence, the angle NRP is equal to the angle NRM; but the angle ARQ is equal to its opposite angle NRM; hence, the lines PR and QR make equal angles with AK; consequently they are the required lines.

Prop. XVI.—*Show that the sum of the lines drawn from two given points to any point of a given line, is the least possible when these lines are equally inclined to the given line.*

Demonstration.—Employing the same construction as in the last figure, let P and Q be the given points, and PR and QR equally inclined to AK; also PS and SQ unequally inclined.

The line AK is drawn through the vertex R of the triangle QRP and perpendicular to the bisectrix of the angle QRP; hence, from Prop. XII, Key, QR + RP < QS + SP, *which was to be proved.*

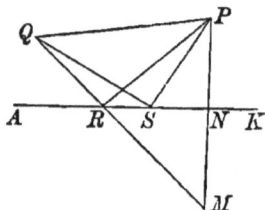

PROP. XVII.—*From two given points, on the same side of a given line, draw two lines meeting on the given line, and equal to each other.*

SOLUTION.—Let A and C be the given points and BD the given line.

Draw AC, and bisect it by a perpendicular PQ; from the point Q, in which this meets the given line, draw QA and QC.

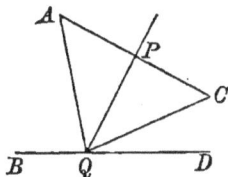

Because PQ is perpendicular to AC, at its middle point, Q is equidistant from A and C (Bk. I. Prop. 16); hence, AC and CQ are the required lines.

PROP. XVIII.—*From a given point A, draw a line that shall be equally distant from two given points B and C.*

SOLUTION.—Let A be the first point, and let B and C be the other given points.

Draw BC, bisect it in O, and then draw AO. From B and C, draw BQ and CP perpendicular to AO.

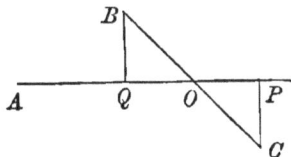

The triangles BOQ and COP are equal in all their parts; hence BQ equals CP. The line AO is therefore the required line.

PROP. XIX.—*Through a given point draw a line cutting the sides of a given angle and making the interior angles equal to each other.*

SOLUTION.—Let P be the given point and DAF the given angle.

Draw the bisectrix AS, of the given angle, and through P draw PR perpendicular to AS, cutting the sides of the angle in Q and R.

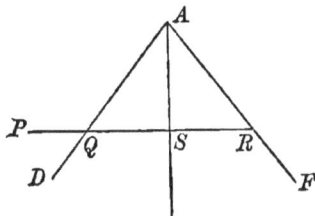

In the triangles ASQ and ASR, we have, AS common, the angles SAQ and

SAR eqnal by construction, and the angles ASQ and ASR equal, because they are right angles; hence, these triangles are equal in all their parts. The angles SQA and SRA, are therefore equal, and consequently PR is the required line.

PROP. XX.—*Draw a line* PQ *parallel to the base* BC *of a triangle* ABC, *so that* PQ *shall be equal to the sum of* BP *and* CQ.

SOLUTION.—Let ABC be the given triangle.

Draw the bisectrices BR and CR, of the angles B and C, and through their intersection R, draw PQ parallel to BC.

Since PQ is parallel to BC, the angles QRC and RCB are equal; but, RCQ and RCB are equal by construction; hence, the angles QRC and QCR are equal; the triangle RQC is therefore isosceles (Bk. II. Prop. 12), and consequently RQ is equal to QC. In like manner it may be shown that RP is equal to PB. Hence,

$$PR + RQ \text{ or } PQ = BP + QC;$$

the line PQ is therefore the required line.

PROP. XXI.—*In a given isosceles triangle draw a line that shall cut off a trapezoid whose base is the base of the given triangle and whose other sides shall be equal to each other.*

SOLUTION.—Let ADF be the given isosceles triangle.

Draw the bisectrix FP of the angle F, and through the point in which it meets AD draw PQ parallel to DF.

It may be shown as in the last problem, that the triangle PQF is isosceles, the side

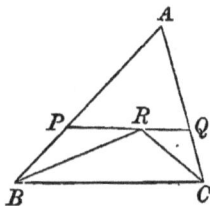

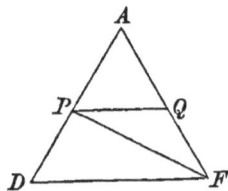

PQ being equal to QF. Since PQ is parallel to DF, the angles AQP
and APQ are equal, and consequently AP is equal to AQ; AD — AP
or PD, is equal to AF — AQ or QF.   Hence, PQ = QF = PD,
and consequently PQ is the required line.

PROP. XXII.—*If two opposite sides of a parallelogram are bisected
and lines be drawn from the points of bisection to the vertices of the
opposite angles, show that these lines will divide the diagonal which
they intersect, into three equal parts.*

DEMONSTRATION.—Let ABCD be any parallelogram, and let P and
Q be the middle points of BC and AD.

Draw PD and BQ, also draw QT
parallel to AC.

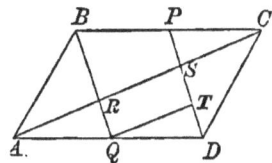

Since BP is parallel and equal to QD,
BQ and PD are also parallel and equal
(Bk. I. Prop. 30).   In the triangles ARQ
and CSP, the angles ARQ and CSP are equal because they are alternate exterior angles with respect to BQ and PD.   The angles RAQ
and SCP are equal, because they are alternate angles, and the sides
AQ and CP are equal because they are halves of equal lines (Ax. 7);
hence, the triangles are equal in all their parts : AR is therefore
equal to SC.   In the triangles ARQ and QTD, the angles ARQ and
QTD are equal because their sides are parallel and lie in the same
direction (Bk. I. Prop. 24); the angles RAQ and TQD are equal
because they are opposite exterior and interior angles, with respect to
BQ and PD, and the sides AQ and QD are equal by hypothesis;
hence the triangles are equal in all their parts : QT is therefore equal
to AR : but QT is equal to RS; hence, QT = AR = SC, *which was
to be .proved.*

PROP. XXIII.—*Construct a triangle, having given the two angles at the base and the sum of the three sides.*

SOLUTION.—Let AB be equal to the sum of the three sides, and let the angles BAC and ABC be equal to the angles at the base of the required triangle.

Draw the bisectrices AT and BT, of the angles A and B, and through their point of intersection T, draw TR parallel to CA or TS parallel to CB.

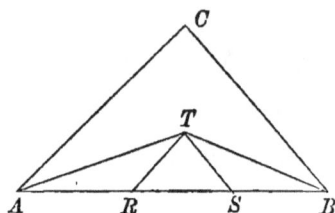

The angles RTA and TAC are equal because they are alternate; but TAC is equal to ATR by construction; hence, the triangle ART is isosceles, and consequently AR = RT. In like manner BS = ST. The angle TRS is equal to CAS, the angle TSR is equal to CBA and the sum of RT, RS, and ST is equal to AB; hence, TRS is the required triangle.

PROP. XXIV.—*Construct a triangle, having given one angle, one of its including sides and the sum of the other two sides.*

SOLUTION.—Let ABC be the given angle, AB one of its including sides, and let BD be equal to the sum of the other two sides.

Draw AD. At A make the angle DAC equal to the angle ADC (Bk. III. Prob. 4).

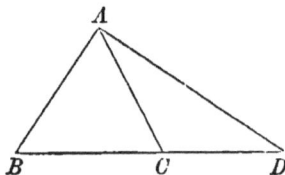

The triangle ACD is isosceles and consequently, AC = CD (Bk. I. Prop. 12). Hence, BC + CA = BC + CD = BD. The triangle ABC is therefore the required triangle.

PROP. XXV.—*Construct an equilateral triangle, having given one of its altitudes.*

SOLUTION.—Let AD be equal to the given altitude. Draw any line AQ, perpendicular to AD, and on any part of it, as AB, construct an equilateral triangle ABC. Through D, draw DP parallel to AQ, intersecting AC, produced in P. Through P draw PQ parallel to CB: then will APQ be the required triangle.

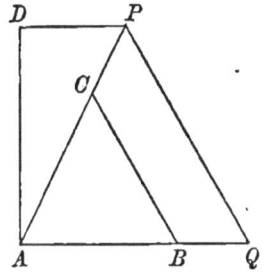

For, the triangles ACB and APQ, being similar, and ACB equilateral, APQ is also equilateral. But the altitude of the triangle APQ is equal to the given altitude: hence, APQ is the required triangle.

PROP. XXVI.—*Show that the three altitudes. of a triangle all intersect in a common point.*

DEMONSTRATION.—Let ABC be the given triangle.

Through A, draw PQ parallel to CB, through B, draw RQ parallel to CA, and through C, draw PR parallel to AB. Through A, B and C, draw perpendiculars to PQ, QR, and RP, meeting the sides of the given triangle in G, F, and E.

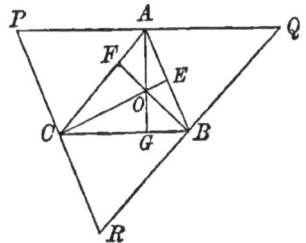

Now, BQ and CA are equal, because they are parallels between parallels; for a like reason RB and AC are equal: hence, B is the middle point of QR. In like manner, C is the middle point of RP, and A is the middle point of PQ. Because PQ, QR, and RP are chords of the circle that circumscribes the triangle PQR, the perpendiculars, AG, BF, and CE, pass through the centre O of that circle : but these perpendiculars are also the altitudes of the given triangle; hence, these altitudes all intersect at the same point, *which was to be proved.*

PROP. XXVII.—*If one of the acute angles of a right-angled triangle is double the other, show that the hypothenuse is double the smaller side about the right angle.*

DEMONSTRATION.—Let ABC be a right-angled triangle in which the angle C is double the angle A.

Draw the bisectrix CO of the angle C, and from the point O in which it meets BA, draw OP perpendicular to the hypothenuse CA.

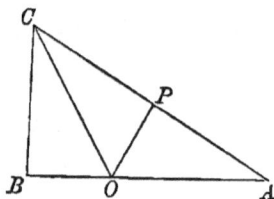

Because the angle C is equal to twice the angle A, we have the angle OCA equal to the angle OAC; that is, the triangle OCA is isosceles; hence, the perpendicular OP, bisects the side CA; consequently CP is equal to one-half of the hypothenuse. But CP is equal to CB, because the right-angled triangles CPO and CBO have the same hypothenuse CO, and the acute angle OCP in the one, equal to the acute angle OCB in the other. Hence, CA = 2CP = 2CB, *which was to be proved.*

PROP. XXVIII.—*Let a median be drawn from the vertex of any angle A of a triangle ABC; then show that the angle A is a right angle when the median is equal to half the side BC, an acute angle when the median is greater than half of BC, and an obtuse angle when it is less than half of BC.*

DEMONSTRATION.—Let ABC be a triangle, and let AD be the median drawn from the vertex A.

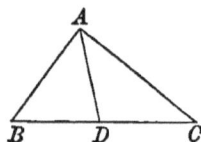

1°. Suppose AD = BD = DC.

The triangle ADC is isosceles, because its sides AD and DC are equal; hence, the angles DAC and DCA are equal. In like manner it may be shown that the angles DAB and DBA are equal. Hence, the sum of the angles CAD and DAB, or the angle BAC, is equal to the sum of the angles ABD and ACD, that is, the angle at A is equal to half the sum of the three angles of the given triangle, or to one right angle, *which was to be proved.*

2°. Suppose AD > BD and consequently AD > CD.

Since, AD > BD, we have from the converse of Prop. 9, Bk. I, the angle B greater than the angle BAD. In like manner the angle C can be shown to be greater than the angle DAC. Hence, the sum of the angles BAD and DAC, or the angle at A, is less than half the sum of the three angles of the given triangles, and consequently it is acute, *which was to be proved.*

3°. Suppose AD < BD, and consequently AD < DC.

It follows from the converse of Prop. 9, Bk. I, that the angle B is less than the angle BAD; and the angle C is less than the angle CAD; hence, the angle at A is greater than half the sum of the three angles of the given triangle, that is, it is obtuse, *which was to be proved.*

PROP. XXIX.—*Let any quadrilateral be inscribed about a circle : then show that the sum of two opposite sides is equal to the sum of the other two opposite sides.*

DEMONSTRATION.—Let ABCD be any quadrilateral circumscribed about the circle O, and let its sides be tangent to this circle at the points P, Q, R, and S.

From the principle demonstrated in the
Corollary to Prob. 14, Bk. III, we have the
following equations:

$$DS = DP$$
$$CS = CR$$
$$AQ = AP$$
$$BQ = BR.$$

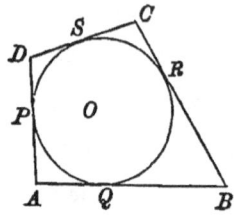

Adding these equations, member to member, and recollecting that
DS + CS = DC, AQ + BQ = AB, DP + AP = DA, and
CR + BR = CB, we have,

$$DC + AB = DA + CB,$$

which was to be proved.

Prop. XXX.—*Draw a straight line tangent to two given circles.*

Solution.—Let A and B be the centres of the two circles, and
let AP and BQ be their radii, and
let AP > BQ.

Draw the line AB, through their
centres.

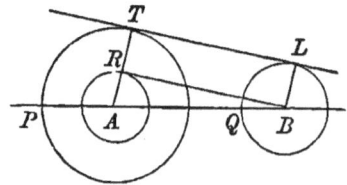

With a radius AR, equal to
AP — BQ, and a centre A, de-
scribe the circle AR; through B draw BR tangent to this auxiliary
circle, and let AR be drawn through the point of contact R, and
prolonged to T; draw the radius BL, parallel to AT, and through the
points T and L draw the line TL.

Because AR is perpendicular to BR (Bk. III, Prop. 9), and BL
is parallel to AR, the angles TRB and LBR are right angles. But,
RT is equal to BL, by construction; hence, the figure TRBL is a
rectangle, and consequently the line TL is perpendicular to the radii

AT and BL, at their extremities; TL is therefore tangent to both the given circles; hence, TL is the required line.

Scholium.—A second tangent may, in like manner, be constructed on the other side of the circles, and it may be shown that both intersect the line of centres at a common point, to the right of B.

By using an auxiliary circle whose radius is equal to AP + BQ, two other tangents may be constructed intersecting each other on the line of centres, between A and B. The first pair of tangents is said to be external, the second pair is said to be internal.

There may be several positions of the given circles.

1°. If the circles are external to each other, they will always have two external and two internal tangents, common to both.

2°. If the circles touch each other externally, they will have two external tangents and only one internal tangent.

3°. If the circles cut each other, they will have two external tangents but no internal tangent.

4°. If the circles touch each other internally, they will have one external but no internal tangent.

5°. If one circle lies wholly within the other, they can have no common tangent whatever.

Prop. XXXI.—*Through a given point* P, *draw a circle that shall be tangent to a given line* CB, *at a given point* B.

Solution.—Let P be a given point, CB a given line, and B a given point on that line.

Draw PB and bisect it by a perpendicular OC, meeting BC in C; at B erect BO perpendicular to CB, and meeting OC in O. Then, with O as a centre, and OB as a radius describe a circle.

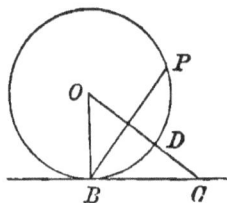

Because OD is perpendicular to PB, at its middle point, O is equally distant from B and P (Bk. I, Prop. 16) ; hence, the circle passes through P. Because BC is perpendicular to the radius OB, at its extremity, it is tangent to the circle at B. Hence, the circle BDP is the required circle.

PROP. XXXII.—*Let two circles intersect each other, and through either point of intersection let diameters of the circles be drawn : then show that the other extremities of these diameters and the other point of intersection, lie in the same straight line.*

DEMONSTRATION.—Let C and D be the centres of two circles which intersect each other at E and S, and let EQ and EP be the diameters of the circles through E.

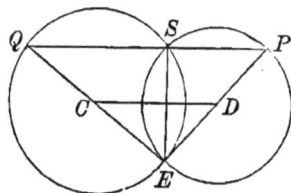

Draw QS, PS, and SE.

Because ECQ is a diameter of the circle C, the angle QSE is a right angle (Bk. III, Prop. 18, Cor. 2); in like manner the angle ESP is a right angle : hence, the line ES meets QS and SP, making the sum of the angles ESP and ESQ equal to two right angles; the lines QS and SP therefore form a single straight line QSP (Bk. I, Prop. 3), *which was to be proved.*

PROP. XXXIII.—*Through two given points, A and B, draw a circle that shall be tangent to a given line CP.*

SOLUTION.—Let A and B be the given points and CP the given line.

Draw AB, and prolong it till it intersects CP at C; lay off BQ equal to AC, and on CQ as a diameter, construct a semicircle QDC; draw BD perpendicular to CQ, cutting this semi-

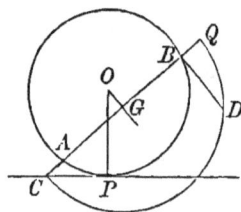

circle in D; make CP = BD and draw PO perpendicular to CP; also draw GO perpendicular to, and bisecting AB; from the point O, in which the last two lines intersect, as a centre, and with a radius equal to OP draw a circle BPA.

The circle BPA is tangent to the line PC, at P, because CP is perpendicular to the radius OP, at its extremity. Since BD is perpendicular to CQ, it is a mean proportional between CB and BQ, that is; between CB and CA (Bk. IV, Prop. 23); hence, the points A and B are on the circle APB (converse of Prop. 30, Bk. IV): consequently the circle BPA is the required circle.

PROP. XXXIV.—*Draw a circle that shall be tangent to a given circle C and also to a given line* BP, *at a given point* P.

SOLUTION.—Let C be the given circle, BP the given line, and P the given point.

Through P draw the line QO, perpendicular to BP, and make PQ equal to the radius CD of the circle C; draw QC, and bisect it by a perpendicular SO; from the point O, in which SO and QO intersect, as a centre, and with a radius OP, describe the circle PD.

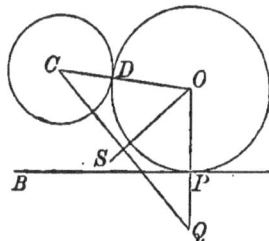

Because SO is perpendicular to QC, at its middle point, OQ = OC and consequently OP = OD. But BP is tangent to the circle O, at P, because it is perpendicular to the radius OP at its extremity. The circle O is tangent to the circle C, because the distance OC = OD + DC (Bk. III, Prop. 14, Scho.). Hence, the circle O is the required circle.

PROP. XXXV.—*Draw a circle that shall be tangent to a given line,* TP, *and also to a given circle* C, *at a given point* Q.

SOLUTION.—Let TP be the given line, C the centre of the given circle, and let Q be any given point on the circle C.

Through Q draw the line CQO, and through the same point Q draw QT perpendicular to CO; draw the bisectrix of the angle QTP, meeting CO in O; draw OP perpendicular to TP, and then with O as a centre and with OP as a radius, describe the circle O.

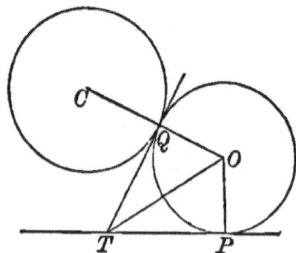

Since O is on the bisectrix TO, OP and OQ are equal; hence, the circle whose radius is OP, passes through Q, and is tangent to the circle C at Q; it is also tangent to TP; hence, it is the required circle.

PROP. XXXVI.—*Draw a circle that shall pass through a given point* Q, *and be tangent to a given circle* C, *at a given point* P.

SOLUTION.—Let Q be the given point, C the centre of the given circle, and P the given point on that circle.

Draw PQ, and bisect it by the perpendicular DO; draw CP, and prolong it till it meets DO, in O; with O as a centre and OP as a radius, describe the circle OP.

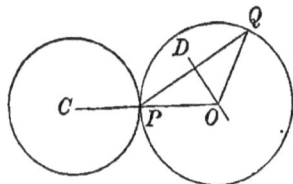

Because DO is perpendicular to PQ, at its middle point, OQ = OP; hence, the circle OP passes through Q. Because CO = CP + PO the circle OP is tangent to the circle C. Hence, the circle OP is the required circle.

PROP. XXXVII.—*Draw a circle, with a given radius, that shall be tangent to a given line DP, and to a given circle C.*

SOLUTION.—Let DB be the given radius, DP the given line, and C the centre of the given circle, and CT its radius.

Draw a line BO, parallel to DP, and at a distance from it equal to DB. With C as a centre, and a radius CO, equal to CT + DB, describe an arc cutting BO in the points B and O. With B as a centre, and with a radius equal to BD, describe a circle; also with O as a centre, and the same radius, describe a second circle.

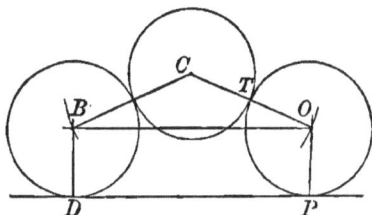

These circles are tangent to the line DP, at the points D and P; they are also tangent to the given circle, because the distance from C, to the centre of either, is equal to the sum of the radii CT and BD; hence, either of the circles B and O, is the required circle.

PROP. XXXVIII.—*Find a point, in the prolongation of any diameter of a given circle, such that a tangent from it to the circumference shall be equal to the diameter of the circle.*

SOLUTION.—Let C be the centre of the given circle, and let AB be any diameter.

At B, draw BP perpendicular to AB and equal to the diameter AB; draw PC, and at the point Q, in which it cuts the circumference, draw the tangent QT and prolong it till it meets the prolongation of AB, in T.

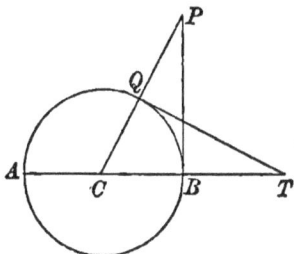

The triangles CQT and CBP have the common angle PCT; the angles CQT and CBP are equal, because they are right angles; and the sides CQ and CB are equal, because they are radii of the same circle: hence, these triangles are equal in all their parts, and consequently, QT = BP = AB; T is therefore the required point.

PROP. XXXIX.—*Show that when two circles intersect each other, the longest common secant that can be drawn through either point of intersection is parallel to the line joining the centres of the circles.*

DEMONSTRATION.—Let C and D be the centres of two circles intersecting each other at S and R, and let KT be a common secant drawn through one of the points of intersection S.

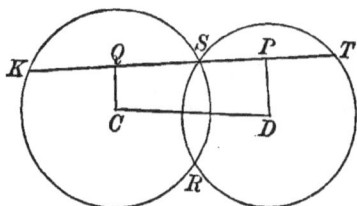

Draw CD; and from C and D let fall CQ and DP perpendicular to KT, and meeting it in the points Q and P.

The distances KQ and QS are equal (Bk. III, Prop. 6), as are also the distances SP and PT; hence, KT = 2QP: but QP is the projection of CD on the line KT, and this projection is the greatest possible when KT is parallel to CD (Prop. VI, Key). Hence, the whole line KT is the greatest possible when it is parallel to CD, *which was to be proved.*

PROP. XL. PROBLEM.—*Construct the greatest possible equilateral triangle whose sides shall pass through three given points, A, B, and C, not in the same straight line.*

SOLUTION.—Let A, B, and C be the three given points.

Join the three points, forming the triangle ABC; on the sides AB and AC, as chords, construct segments of circles APB and AQC capa-

ble of containing an angle equal to two-thirds of a right angle
(Bk. III, Prob. 16); through A draw QAP
parallel to the line through D and E, the
centres of the segments; from the points Q
and P, draw QC and PB, and prolong them
till they meet at R.

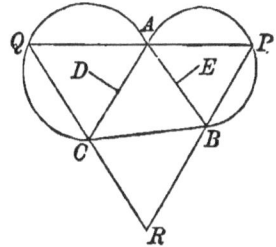

Any triangle whose sides pass through
A, B, and C, and whose two vertices are in
the arcs APB and AQC, is equilateral, because each of its angles is
equal to two-thirds of a right angle; but no other triangle whose sides
pass through A, B, and C, can be equilateral. Now, of the group of
equilateral triangles that can be formed, as just indicated, that one
will be the greatest possible which has one of its sides the greatest
possible; but from Prop. XXXIX, Key, the side through A will be
greatest possible when it is parallel to the line joining D and E:
hence, QPR is the required triangle.

PROP. XLI.—*Show that the bisectrices of the four angles of any
quadrilateral intersect in four points, all of which lie on the circumference of the same circle.*

DEMONSTRATION—Let ABCD be any quadrilateral, and let AR,
BR, CP and DP, be the bisectrices
of its four angles. Draw PR.

The angle PSR, or its equal ASD,
is equal to one-half the sum of ABC
and DCB (Prop. III, Key). In like
manner it may be shown that the
angle PQR is equal to one-half the sum of the angles BAD and CDA.
Hence, the sum of the angles PSR and PQR is equal to one-half the

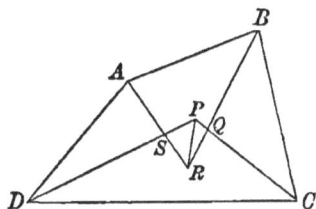

sum of the four angles of the given quadrilateral, that is, to two right angles. Consequently, from the converse of Prop. 18, Bk. III, Cor. 4, the quadrilateral SPQR can be inscribed in a circle, *which was to be proved.*

PROP. XLII.—*If two circles touch each other externally, and if two common secants are drawn through the point of contact and terminating in the concave arcs, show that the lines joining the extremities of these secants, in the two circles, are parallel.*

DEMONSTRATION.—Let O and P be the centres of two circles which are tangent to each other at the point Q.

Draw the secants BC and DE, terminating in the concave arcs BD and EC; draw also BD and EC, and at Q draw TS tangent to both circles.

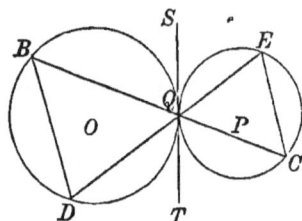

The angle SQE is equal to QCE (Bk. III, Prop. 21 and 18); in like manner the angle TQD is equal to QBD. But SQE and TQD are equal because they are opposite angles; hence, QCE and QBD are equal, and consequently, the lines CE and BD are parallel (Bk. I, Prop. 19, Cor. 1); *which was to be proved.*

PROP. XLIII.—*Let an equilateral triangle be inscribed in a circle, and let two of the subtended arcs be bisected by a chord: then show that the sides of the triangle divide the chord into three equal parts.*

DEMONSTRATION.—Let ABC be an equilateral triangle inscribed in a circle ASBCP, and let P and S be the middle points of the arcs APC and ASB.

Draw PS, cutting AC and AB in Q and R; also draw AP and AS.

The angles ASR and RAS are equal because they are measured by halves of the
equal arcs AP and SB; hence, AR = RS.
In like manner it may be shown that AQ =
PQ. In the triangle AQR, the angle ARQ
is measured by $\frac{1}{2}$ (SB + PA), the angle
AQR is measured by $\frac{1}{2}$ (AS + PC), and the angle QAR is measured
by $\frac{1}{2}$CB, hence, the three angles are equal, and consequently QR =
AQ = AR, or PQ = QR = RS, *which was to be proved.*

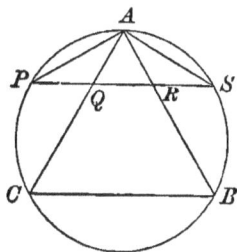

PROP. XLIV.—*Find a point within a triangle, such that the angles
formed by drawing lines from it to the three vertices of the triangle,
shall be equal.*

SOLUTION.—Let ABD be the given triangle. On AB construct
an arc that will contain an angle equal to
four-thirds of a right angle; on BD construct another arc that will contain an
angle equal to four-thirds of a right angle,
and intersecting the first at O; draw OD,
OA, and OB.

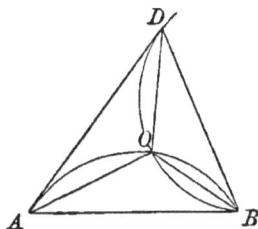

The angles AOD and DOB are each
equal to four-thirds of a right angle, and consequently, the angle AOB
is also equal to four-thirds of a right angle; hence, O is the required
point.

PROP. XLV.—*Inscribe a circle in a quadrant of a given circle.*

SOLUTION.—Let CBPD be the given quadrant.

Draw the bisectrix CP of the angle DCB, and at P draw a tangent
PT to the quadrant, meeting CB pro-
duced at T.

Draw the bisectrix TO, of the
angle CTP, and from the point O, in
which it meets CP, draw OR and OQ
perpendicular to CB and CD; with
O as a centre, and OR as a radius,
describe a circle.

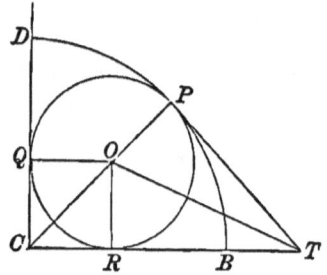

Because TO is the bisectrix of the angle RTP, OR = OP
(Prop. IV, Key); because CP is the bisectrix of the angle BCD,
OR = OQ: hence, the circle OR passes through P and Q. This
circle is tangent to CT because CT is perpendicular to OR at R, and
for a like reason it is also tangent to CD at Q; the circle OR is tangent to the arc DPB, because CO = CP — OP: hence, the circle
OR is the required circle.

Prop. XLVI.—*Through a given point* P, *within a given angle*
ABC, *draw a circle that shall be tangent to both sides of that angle.*

Solution.—Let P be the given point and ABC the given angle.

Draw the bisectrix BO, of the
given angle, and also the line BP;
from any point Q, of BO, draw QR
perpendicular to AB, and from Q,
as a centre, and with a radius QR,
draw the arc RS, cutting BP in S;
draw SQ and also draw PO parallel
to SQ, intersecting BO at O; then draw OA parallel to QR, and with
O as a centre and OA as a radius, describe the circle APC: also draw
OC perpendicular to BC.

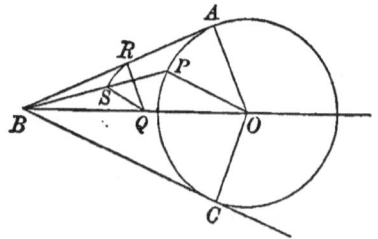

The triangles BSQ and BPO have the angle at B common, and since QS is parallel to OP, the angles BQS and BOP are equal, as are also the angles BSQ and BPO; hence, the triangles are similar (Bk. IV, Prop. 18), and consequently, their corresponding sides are proportional.

We have, therefore, the proportion

$$BQ : BO :: QS : OP \quad . \quad . \quad . \quad (1).$$

In like manner, we have from the triangles BQR and BOA, the proportion,

$$BQ : BO :: QR : OA \quad . \quad . \quad . \quad (2).$$

Because the first three terms of proportions (1) and (2) are equal, each to each, their fourth terms must be equal, that is, OP = OA. Hence, the circle whose radius is OA, passes through P. The distance OC is equal to OA (Prop. IV, Key); hence, the circle whose centre is O passes through C. Furthermore, BA is perpendicular to OA, at its extremity, and BC is perpendicular to OC, at its extremity; hence, the circle whose radius is OA, is tangent to both sides of the given angle; it is therefore the required circle.

PROP. XLVII.—*Show that the middle points of the sides of any quadrilateral are the vertices of an inscribed parallelogram.*

DEMONSTRATION.—Let ABCD be any quadrilateral, and let P, Q, R, and S, be the middle points of its four sides.

Draw PQ, QR, RS, and SP; also draw the diagonals BD and AC.

Because AQ = QB, and AP = PD, we have the proportion,

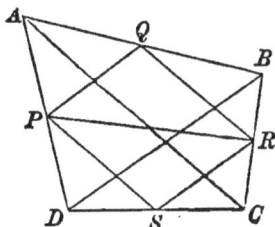

$$AP : PD :: AQ : QB.$$

Hence, from Bk. IV, Prop. 16, PQ is parallel to DB. In like manner it may be shown that RS is parallel to DB: hence, PQ and SR are parallel to each other. The lines QR and PS are parallel to AC, for similar reasons; hence, they are parallel to each other, and consequently, the figure PQRS is a parallelogram inscribed in the given quadrilateral, *which was to be proved.*

PROP. XLVIII.—*Inscribe in a given triangle, a triangle whose sides shall be parallel to the sides of a second given triangle.*

SOLUTION.—Let ABC be the first given triangle, and let DEF be the second given triangle.

Take any point P, on the side AB, and through it draw PQ parallel to DE; from Q draw QR parallel to EF, and from P draw PR parallel to DF, intersecting PR in R; draw BR and prolong it to S. Through S draw SM parallel to RQ, and SN parallel to RP, and then join M and N.

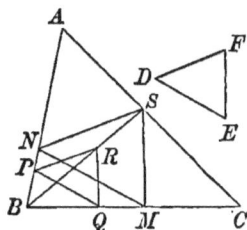

From the triangles BSN and BRP, we have (Bk. IV, Prop. 15),

$$BR : RS :: BP : PN \quad . \quad . \quad . \quad (1).$$

From the triangles BSM and BRQ, we have, like reasons,

$$BR : RS :: BQ : QM \quad . \quad . \quad . \quad (2).$$

Combining proportions (1) and (2), we have,

$$BP : PN :: BQ : QM;$$

hence, from (Bk. IV. Prop. 16), we have, NM parallel to PQ. The sides of the inscribed triangle NSM, are therefore parallel to the sides of the triangle PRQ, and consequently, to the sides of the triangle DFE. The triangle NSM is therefore the required triangle.

PROP. XLIX.—*Through a point* P, *within a given angle, draw a line such that it and the parts of the sides that are intercepted shall contain a given area.*

SOLUTION.—Let BAC be the given angle, and P the given point.

Through P, draw DPF parallel to AC, and complete the parallelogram DFGA, in such manner that its area shall be equal to the given area. To do this, first construct a triangle equal to the given area (Bk. IV, Prob. 6), then convert this into an equivalent triangle whose altitude is equal to the altitude DE, and then make DF and AG each equal to one-half the base of the last triangle. From G, draw GH perpendicular to AC and make it equal to PD; with H as a centre, and a radius equal to PF describe an arc cutting AC in C; then draw CP and prolong it to B.

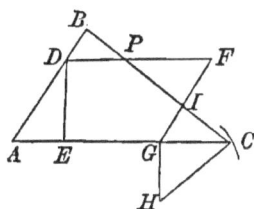

The triangles DBP, PFI, and IGC are similar (Bk. IV, Prop. 21); they are therefore proportional to the squares of their homologous sides DP, PF, and GC; hence, we have the continued proportion

$$\text{PFI} : \text{DBP} : \text{GIC} :: \overline{\text{PF}}^2 : \overline{\text{PD}}^2 : \overline{\text{GC}}^2 \quad . \quad . \quad . \quad (1),$$

whence, by composition, (Bk. II, Prop. 6),

$$\text{PFI} : \text{DBP} + \text{GIC} :: \overline{\text{PF}}^2 : \overline{\text{PD}}^2 + \overline{\text{GC}}^2 \quad . \quad . \quad . \quad (2).$$

But, from the nature of the construction (Bk. IV, Prop. 11), we have,

$$\overline{\text{HC}}^2 \text{ or } \overline{\text{PF}}^2 = \overline{\text{HG}}^2 + \overline{\text{GC}}^2 = \overline{\text{PD}}^2 + \overline{\text{GC}}^2;$$

hence, the terms of the second couplet of proportion (2) are equal, and consequently the terms of the first couplet are equal, that is,

$$\text{PFI} = \text{DBP} + \text{GIC}.$$

If now we take away from the parallelogram AF, the triangle PFI and then add to it its equal, DBP + GIC, we shall have the triangle BAC. The triangle BAC is therefore equal to the parallelogram AF, which is equal to the given area; hence, BAC is the required triangle.

PROP. L.—*Construct a parallelogram, whose area and perimeter are respectively equal to the area and perimeter of a given triangle.*

SOLUTION.—Let ABC be the given triangle. Prolong AB, making BD = BC, and bisect AD in E; draw BF parallel to AC; with A as a centre and a radius equal to AE, describe an arc cutting BF in G; draw AG; bisect AC, in II, and complete the parallelogram AHFG.

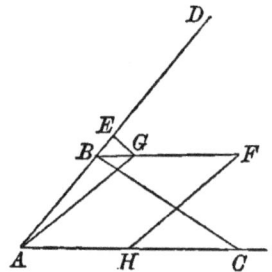

The sides AG and HF of the parallelogram AF, are each equal to AE, and consequently their sum is equal to AD, or AB + BC. The sides AH and GF are each equal to one-half of AC, and consequently their sum is equal to AC; hence, the perimeter of the parallelogram AF is equal to that of the given triangle. The altitude of the parallelogram is equal to that of the given triangle, and the base of the parallelogram is one-half that of the given triangle; hence, the area of the parallelogram is equal to that of the given triangle (Bk. IV, Prop. 5 and 6). The parallelogram AGFII is therefore the required parallelogram.

PROP. LI.—*Inscribe a square in a semicircle, that is, a square, two of whose vertices are in the diameter and the other two in the semicircumference.*

SOLUTION.—Let ADCB be a semicircumference whose diameter is AB, and whose centre is O.

Draw AE perpendicular to, and equal to, AB; draw EO, cutting the semicircumference in D; draw DC parallel to AB; also, draw DF and CG perpendicular to AB; draw OC.

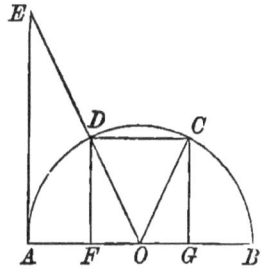

The triangles OAE and OFD are similar; but, OA = $\frac{1}{2}$AE, and consequently OF = $\frac{1}{2}$FD. The triangles OFD and OGC are equal in all their parts; hence, OG = OF, and consequently FG = FD; but FD = GC, and FG = DC; hence, FDCG is the required square.

Prop. LII.—*Through a given point P, draw a line cutting a triangle, so that the sum of the perpendiculars to it from the two vertices, on one side of the line, shall be equal to the perpendicular to it, from the vertex, on the other side of the line.*

Solution.—Let P be the given point and ABC the given triangle.

Draw the median AD, and on it lay off DO = $\frac{1}{3}$DA; draw PO, and prolong it till it meets DC prolonged, in S; upon this line let fall the perpendiculars CG, AF, DK and BE.

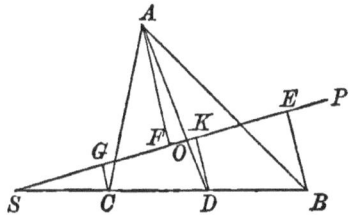

Because D is the middle of CB, we have the following equations:

$$SD = SC + \tfrac{1}{2}CB$$
$$SD = SB - \tfrac{1}{2}CB.$$

Adding, and dividing by 2, we have,

$$SD = \tfrac{1}{2}(SC + SB) \quad . \quad . \quad . \quad (1).$$

From the similar triangles SCG and SBE, we have,

$$SC : SB :: CG : BE \quad . \quad . \quad . \quad (2);$$

whence, by composition

$$SC : SC + SB :: CG : CG + BE \quad . \quad . \quad . \quad (3).$$

From the similar triangles SCG and SDK, we have

$$SC : SD :: CG : DK \quad . \quad . \quad . \quad (4).$$

Because the antecedents in (3) and (4) are equal, each to each, we have,

$$SC + SB : SD :: CG + BE : DK \quad . \quad . \quad . \quad (5).$$

But, SC + SB = 2SD, hence,

$$CG + BE = 2DK \quad . \quad . \quad . \quad (6).$$

From the similar triangles AFO and DKO, we have,

$$AF : DK :: AO : DO \quad . \quad . \quad . \quad (7);$$

but, AO = 2DO by construction, hence,

$$AF = 2DK \quad . \quad . \quad . \quad (8).$$

From equations (6) and (8), we have,

$$AF = CG + BE;$$

hence, PS is the required line.

PROP. LIII.—*Show that the line which joins the middle points of two opposite sides of any quadrilateral bisects the line joining the middle points of the two diagonals.*

DEMONSTRATION.—Let ABCD be any quadrilateral, F and H the middle points of two opposite sides; and let L and K be the middle points of the two diagonals BD and AC.

Draw LK, FH, FL and KH.

Because F is the middle point of AD, and L the middle point of BD, FL is

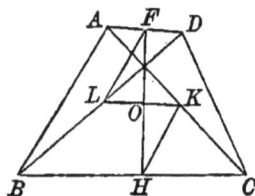

parallel to AB, and equal to ½AB.   In like manner it may be shown, that HK is parallel to AB, and equal to ½AB.   Hence, HK and LF are parallel and equal to each other.   The triangles FOL and HOK, have their angles equal each to each and the sides LF and KH equal; hence, they are equal in all their parts, and consequently, OL = OK, *which was to be proved.*

PROP. LIV.—*If from the extremities of one of the oblique sides of a trapezoid, lines be drawn to the middle of the opposite side, show that the triangle thus formed is equal to one-half the given trapezoid.*

DEMONSTRATION.—Let BPQR be a trapezoid, and let BD and PD be lines drawn from B and P, to D, the middle point of QR.

Through D, draw TD parallel to BP, and prolong it till it meets the prolongation of BR, in S.

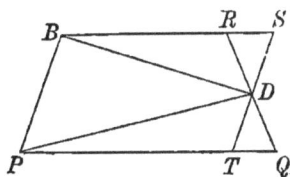

The triangles DRS and DTQ, have their corresponding angles equal and the sides DR and DQ equal; hence, they are equal in all their parts.   If we take the triangle DTQ from the given trapezoid, and then add the equal triangle DRS, we shall have the parallelogram BPTS; hence, this parallelogram is equal to the given trapezoid.   The parallelogram PTSB, and the triangle PBD, have a common base BP and the same altitude; viz., the perpendicular distance from D, to PB; hence, from Prop. 5 and 6, Bk. IV, the triangle is equal in area to one-half the parallelogram, and consequently, to one-half the given trapezoid, *which was to be proved.*

PROP. LV.—*Find a point in the base of a triangle, such that the lines drawn from it parallel to, and limited by, the other sides of the triangle, shall be equal to each other.*

SOLUTION.—Let ABC be the given triangle. Lay off AE equal to AC, and complete the parallelogram ACFE.

Draw FA cutting BC in O; draw OP parallel to FC, and OQ parallel to FE.

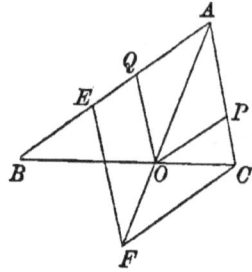

From the similar triangles AFC and AOP, we have,

$$AF : AO :: FC : OP \ldots (1).$$

From the similar triangles AFE and AOQ, we have,

$$AF : AO :: FE : OQ \ldots (2).$$

From (1) and (2),

$$FC : OP :: FE : OQ.$$

But, FC = FE; hence OP = OQ; the point O is therefore the required point.

PROP. LVI.—*Show that the line drawn from the middle point of the base, of any triangle, through the middle point of any line of the triangle parallel to the base, will pass through the opposite vertex, if sufficiently produced.*

DEMONSTRATION.—Let ABC be any triangle, SR a line parallel to the base BC, and let F and P, be the middle points of BC and SR.

Draw PF, and prolong it.

The straight line drawn from A to F, bisects SR (Bk. IV. Prop. 22), that is, it passes through P; but only one straight line can be drawn through F and P, consequently, the line FP, if sufficiently produced, will pass through A, *which was to be proved.*

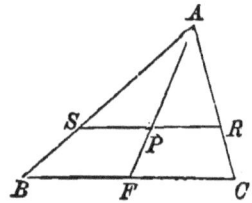

PROP. LVII.—*Show that the three medians of any triangle meet in a common point.*

DEMONSTRATION.—Let ABC be any triangle and let BE and CD be two of its medians, intersecting each other in P.

Draw DE; and through F, the middle point of BC, draw FP, cutting DE in Q, and prolong it.

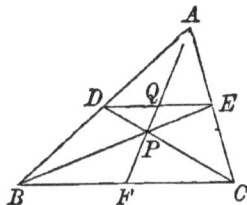

The line DE is parallel to BC; conse-quently the triangles PFB and PQE are similar, as are also the triangles PFC and PQD. From PFC and PQD, we have,

$$PF : PQ :: FC : QD \quad . \quad . \quad . \quad (1).$$

From the triangles PFB and PQE, we have,

$$PF : PQ :: FB : QE \quad . \quad . \quad . \quad (2).$$

From (1) and (2), we have,

$$FC : QD :: FB : QE.$$

But, FC = FB, hence, QD = QE; the line FP must therefore pass through the middle of DE, and therefore, from Prop. LVI, Key, it must also pass through A, that is, the line through A and P is the third median of the given triangle. Hence, the three medians pass through P, *which was to be proved.*

PROP. LVIII.—*On the sides AB and AC of any triangle ABC, construct any two parallelograms ABDE and ACFG; prolong the sides DE and FG till they meet in H; draw HA, and on the third side of the triangle BC, construct a parallelogram two of whose sides are parallel and equal to HA : then show that the parallelogram on BC is equal to the sum of the parallelograms on AB and AC.*

DEMONSTRATION.—Let ABC be any triangle; Let AD and AF be any parallelograms constructed on AB and AC as sides; and let H be the point in which DE and FG meet, when prolonged; let CP be a parallelogram whose sides BP and CQ are parallel and equal to HA.

Draw HA, and prolong it to S; also prolong PB and QC, to K and L.

The parallelograms ABDE and ABKH have a common base AB, and a common altitude; hence, they are equal (Bk. IV, Prop. 1); the parallelograms ABKH and SPBT have equal bases, HA and ST, and a common altitude; they are therefore equal: hence, the parallelogram ABDE is equal to the parallelogram STBP. In like manner it may be shown that the parallelogram ACFG is equal to the parallelogram STCQ. The sum of the parallelograms TP and TQ is equal to the parallelogram BCQP; hence, the parallelogram BCQP is equal to the sum of the parallelograms ABDE and ACFG, *which was to be proved.*

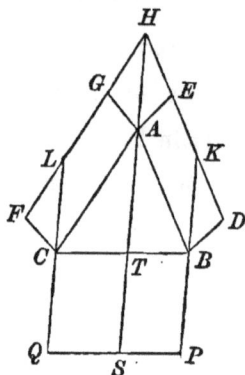

PROP. LIX.—*Assuming the principle demonstrated in the last proposition, deduce from it the truth that the square on the hypothenuse of a right-angled triangle is equal to the sum of the squares on the other two sides.*

DEMONSTRATION.—Let ABC be a triangle right-angled at A, and let BQ, AD, and AF be squares described on its sides. Let DE and FG, when prolonged, meet in H.

Draw HA, and prolong it to S; prolong PB to K, and QC to L.

The right-angled triangle AGH has the side AG $=$ AC, and the side GH = AE = AB; hence, it is equal to the triangle CAB in all its parts, that is, AH is equal to CB, and the angle HAG is equal to the angle ACB. Since the angle CAG is a right angle, the sum of the angles HAG and CAT, or the sum of the angles ACB and CAT is equal to a right angle, and consequently the angle ATC is a right angle; hence, TS is parallel to CQ and BP. The square CP, is therefore a parallelogram whose sides are parallel and equal to HA,

and consequently, from Prop. LVIII, Key, it is equal to the sum of the squares AD and AF, *which was to be proved.*

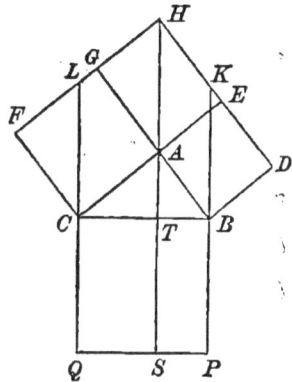

PROP. LX.—*If from the middle of the base of a right-angled triangle a line is drawn perpendicular to the hypothenuse dividing it into two segments, show that the difference of the squares of these segments is equal to the square of the other side about the right angle.*

DEMONSTRATION.—Let ACB be a right-angled triangle, and D the middle of its base.

Draw DE perpendicular to AB; draw also the line AD.

In the right-angled triangle ADE, we have,

$$\overline{AD}^2 = \overline{AE}^2 + \overline{ED}^2 \ldots (1).$$

In the right-angled triangle DEB, we have,

$$\overline{DB}^2, \text{ or } \overline{CD}^2 = \overline{EB}^2 + \overline{ED}^2 \ldots (2).$$

Subtracting (2) from (1), we have,

$$\overline{AD}^2 - \overline{CD}^2 = \overline{AE}^2 - \overline{EB}^2 \ldots (3).$$

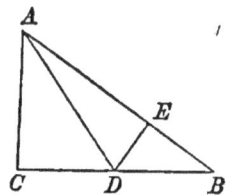

But, $\overline{AD}^2 - \overline{CD}^2 = \overline{AC}^2$, hence, from (3), we have,

$$\overline{AE}^2 - \overline{EB}^2 = \overline{AC}^2,$$

which was to be proved.

PROP. LXI.—*If lines are drawn from any point* P *to the four vertices of a rectangle, show that the sum of the squares of the two lines drawn to the extremities of one diagonal is equal to the sum of the squares of the two lines drawn to the extremities of the other diagonal.*

DEMONSTRATION.—Let ABDC be any rectangle, and P any point in its plane.

Draw PA, PB, PC, and PD; also draw the diagonals AD and CB, and through P draw EPF perpendicular to CD.

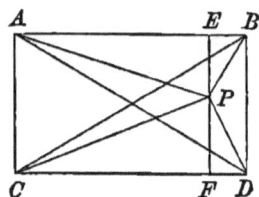

From the right-angled triangles PEB and PFC, we have,

$$\overline{PB}^2 = \overline{PE}^2 + \overline{EB}^2$$

$$\overline{PC}^2 = \overline{PF}^2 + \overline{FC}^2,$$

whence, by addition,

$$\overline{PB}^2 + \overline{PC}^2 = \overline{PE}^2 + \overline{PF}^2 + \overline{EB}^2 + \overline{FC}^2 \ldots (1).$$

From the right-angled triangles PEA and PFD, we have,

$$\overline{PA}^2 = \overline{PE}^2 + \overline{EA}^2$$

$$\overline{PD}^2 = \overline{PF}^2 + \overline{FD}^2,$$

whence, by addition,

$$\overline{PA}^2 + \overline{PD}^2 = \overline{PE}^2 + \overline{PF}^2 + \overline{FD}^2 + \overline{EA}^2 \ldots (2).$$

Because, EB = FD, and EA = FC, the second members of (1) and (2) are equal; and consequently the first members are also equal; hence,

$$\overline{PB}^2 + \overline{PC}^2 = \overline{PA}^2 + \overline{PD}^2,$$

which was to be proved.

PROP. LXII.—*Let a line be drawn from the centre of a circle to any point of any chord ; then show that the square of this line, plus the rectangle of the segments of the chord, is equal to the square of the radius.*

DEMONSTRATION.—Let O be the centre of any circle EBA, BA any chord, and P any point of that chord.

Draw OP, and through P draw the chord ED perpendicular to OP; also, draw EO.

From the right-angled triangle OPE, we have,

$$\overline{OE}^2 = \overline{OP}^2 + \overline{EP}^2 \ldots (1).$$

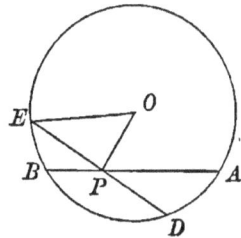

From Prop. 28, Bk. IV, we have,

$$EP \times PD, \text{ or } \overline{EP}^2 = BP \times PA;$$

substituting in (1), we have,

$$\overline{OE}^2 = \overline{OP}^2 + BP \times PA,$$

which was to be proved.

PROP. LXIII.—*Draw a line from the vertex of any scalene triangle to a point in the base, such that this line will be a mean proportional between the segments, into which it divides the base.*

SOLUTION.—Let ABC be a scalene triangle, AC its base, and B its vertex. Through A, B, and C pass a circle ABCG and let its centre be F; draw the diameter BFG, and on the radius FB, as a diameter, describe a second circle cutting AC, in D and R; draw BE, DF and EG.

The angles BDF and BEG, being inscribed in semicircles, are right angles, and consequently, the triangles BDF and BEG are similar; hence, BD = DE.

From Prop. 28, Bk. IV, we have,

$$BD \times DE = AD \times DC, \text{ or } \overline{BD}^2 = AD \times DC;$$

hence, BD is the required line.

A second line could be found, by drawing a line from B to R, that would be a mean proportional between AR and RC.

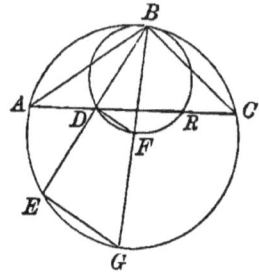

PROP. LXIV.—*Show that the sum of the squares of the diagonals of any quadrilateral is equal to the sum of the squares of the four sides of the quadrilateral, increased by four times the square of the distance between the middle points of the diagonals.*

DEMONSTRATION.—Let ABCD be any quadrilateral, AC and BD its diagonals, and E and F the middle points of the diagonals.

Draw EF, ED, EB, FC and FA.

From Prop. 14, Bk. IV, we have the following relations:

$$\overline{CD}^2 + \overline{CB}^2 = 2\overline{BF}^2 + 2\overline{CF}^2$$

$$\overline{AD}^2 + \overline{AB}^2 = 2\overline{BF}^2 + 2\overline{AF}^2;$$

whence, by addition,

$$\overline{CD}^2 + \overline{CB}^2 + \overline{AD}^2 + \overline{AB}^2 = 4\overline{BF}^2 + 2(\overline{CF}^2 + \overline{AF}^2).$$

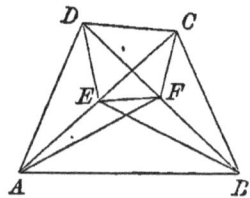

But, from the same proposition, we have,

$$\overline{CF}^2 + \overline{AF}^2 = 2\overline{AE}^2 + 2\overline{EF}^2;$$

which gives, when substituted in the preceding equation,

$$\overline{CD}^2 + \overline{CB}^2 + \overline{AD}^2 + \overline{AB}^2 = 4\overline{BF}^2 + 4\overline{AE}^2 + 4\overline{EF}^2.$$

But, $4\overline{BF}^2 = \overline{BD}^2$, (Bk. IV, Prop. 8, Cor.); also $4\overline{AE}^2 = \overline{AC}^2$; hence,

$$\overline{CD}^2 + \overline{CB}^2 + \overline{AD}^2 + \overline{AB}^2 = \overline{BD}^2 + \overline{AC}^2 + 4\overline{EF}^2,$$

which was to be proved.

Prop. LXV.—*Construct an equilateral triangle equal in area to any isosceles triangle.*

Solution.—Let ABC be any isosceles triangle. On AB, as a side, construct the equilateral triangle ABD; draw the line DCE, which is the common bisectrix of the angles D and C.

On DE, as a diameter, draw a semi-circle DFE, and from C erect the line CF perpendicular to DE; draw EF; and from E, as a centre, with EF as a radius, draw the arc FG, cutting DE in G; through G, draw GP and GQ, parallel to DB and DA. Also draw FD.

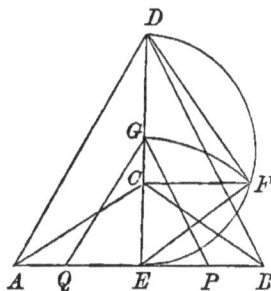

From Prop. 23, Bk. IV, Cor. 2, we have,

$$\overline{EF}^2, \text{ or } \overline{EG}^2 = EC \times ED, \text{ or}$$

$$ED : EG :: EG : EC \quad . \quad . \quad . \quad (1).$$

From the similar triangles EDB and EGP, we have,

$$ED : EG :: EB : EP \quad . \quad . \quad . \quad (2).$$

From (1) and (2), we have,

EG : EC :: EB : EP, or EG × EP = EC × EB;

but, EG × EP is equal to the area of the equilateral triangle QGP, and EC × EB is equal to the area of the isosceles triangle ACB; hence, QGP is the required triangle.

PROP. LXVI.—*In a triangle ABC let two lines be drawn from the extremities of the base BC, intersecting at any point P on the median through A, and meeting the opposite sides in the points E and D; show that DE is parallel to BC.*

DEMONSTRATION.—Let ABC be the given triangle, P any point on the median, and CD, BE, lines drawn from C and B, through P.

Through P, draw SPR parallel to BC; also, join D and E.

From the similar triangles DSP and DBC, we have,

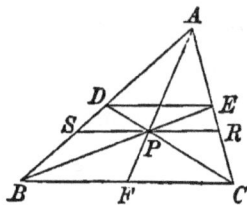

DP : DC :: SP : CB . . . (1).

From the similar triangles EPR and EBC, we have,

EP : EB :: PR, or PS : BC . . . (2).

From (1) and (2), we have,

DP : DC :: EP : EB . . . (3).

Whence, by division (Bk. II, Prop. 6), we have,

DP : DC — DP :: EP : EB — EP, or

DP : PB :: EP : PC.

Hence, the triangles DPE and BPC are similar (Bk. IV, Prop. 20), and consequently, the angles DEP and PBC are equal; hence, DE is parallel to BC, *which was to be proved.*

APPLICATIONS OF ALGEBRA TO GEOMETRY.

PROP. LXVII.—*In a right-angled triangle* ABC, *given the base* BA, *and the sum of the hypothenuse and perpendicular to find the hypothenuse and the perpendicular.*

SOLUTION.—Denote BA by c, BC by x, AC by y, and the sum of BC and AC by s.

Then, $\qquad x + y = s \quad . \quad . \quad . \quad (1).$

From Bk. IV, Prop. 11, $x^2 = y^2 + c^2 \quad . \quad . \quad . \quad (2).$

From (1), we have, $\qquad x = s - y.$

Squaring, $\qquad x^2 = s^2 - 2sy + y^2 \quad . \quad . \quad . \quad (3).$

Subtracting (2) from (3), $0 = s^2 - 2sy - c^2.$

Transposing and dividing, $\quad y = \dfrac{s^2 - c^2}{2s} \; ;$

whence, $\qquad x = s - \dfrac{s^2 - c^2}{2s} = \dfrac{s^2 + c^2}{2s}.$

If $c = 3$ and $s = 9$, we have $x = 5$ and $y = 4$.

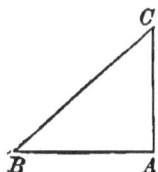

PROP. LXVIII.—*In a right-angled triangle, given the hypothenuse and the sum of the sides about the right angle, to find these two sides.*

SOLUTION.—Denote BC by a, BA by x, CA by y, and BA + AC by s.

Then, $\qquad x + y = s \quad . \quad . \quad . \quad (1)$

and $\qquad x^2 + y^2 = a^2 \quad . \quad . \quad . \quad (2).$

From (1), $\qquad x^2 = s^2 - 2sy + y^2 \quad . \quad . \quad . \quad (3).$

3

Combining (2) and (3),

$$2y^2 - 2sy = a^2 - s^2 \quad \cdots \quad (4),$$

whence,

$$y^2 - sy = \frac{a^2 - s^2}{2} \quad \cdots \quad (5).$$

Solving (5),

$$y = \tfrac{1}{2}s \pm \sqrt{\tfrac{1}{2}a^2 - \tfrac{1}{4}s^2} \quad \cdots \quad (6).$$

Substituting in (1),

$$x = \tfrac{1}{2}s \mp \sqrt{\tfrac{1}{2}a^2 - \tfrac{1}{4}s^2} \quad \cdots \quad (7).$$

If $a = 5$ and $s = 7$, we have, $y = 4$ and 3, $x = 3$ and 4.

PROP. LXIX.—*In a rectangle, given the diagonal and the perimeter, to find the sides.*

SOLUTION.—Denote AC by d, the side BA by x, the side BC by y, and the sum of BA and BC, or half the perimeter by a.

Then,

$$x + y = a,$$

and,

$$x^2 + y^2 = d^2,$$

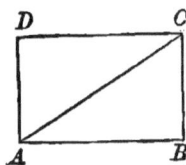

from which we obtain by solution,

$$y = \tfrac{1}{2}a \pm \sqrt{\tfrac{1}{2}d^2 - \tfrac{1}{4}a^2}$$
$$x = \tfrac{1}{2}a \mp \sqrt{\tfrac{1}{2}d^2 - \tfrac{1}{4}a^2}.$$

If $d = 10$ and $a = 14$, we have $y = 8$ and 6, and $x = 6$ and 8.

PROP. LXX.—*Given the base and perpendicular of a triangle, to find the side of an inscribed square.*

SOLUTION.—Let ABC be the given triangle, and let FGHE be the inscribed square.

Denote AB by b, CD by a, GH by x, whence CI is equal to $a - x$.

From the similar triangles ACB and GCF, we have,
$$AB : CD :: GF : CI,$$
or,
$$b : a :: x : a - x,$$
whence,
$$x = \frac{ab}{a + b}.$$

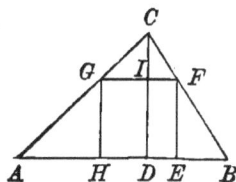

The side of the inscribed square depends only on the base and altitude of the given triangle.

PROP. LXXI.—*In an equilateral triangle, given the distances from a point within to the three sides, to find one side of the triangle.*

SOLUTION.—Let ACB be an equilateral triangle, D any point within it, DE, DG, and DF perpendiculars from D to the sides of the triangle and CH the altitude of the triangle.

Draw DA, DB, and BC.

Denote DG by a, DE by b, and DF by c; also denote one of the sides of the triangle by $2x$, hence, AH $= x$ and CH $= \sqrt{\overline{AC}^2 - \overline{AH}^2} = \sqrt{4x^2 - x^2}$, or CH $= x\sqrt{3}$.

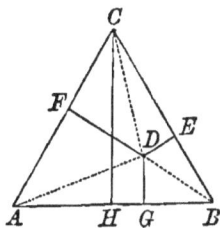

The area of the given triangle is equal to $x \times x\sqrt{3}$, or $x^2\sqrt{3}$;

the area of the triangle ADB is equal to $a \times x$;

the area of the triangle BDC is equal to $b \times x$;

the area of the triangle ADC is equal to $c \times x$.

But the last three triangles make up the first; hence,

$$x^2\sqrt{3} = ax + bx + cx,$$

whence, by solution,
$$x = \frac{a + b + c}{\sqrt{3}}.$$

PROP. LXXII.—*In a right-angled triangle, having given the base and difference between the hypothenuse and perpendicular, to find the sides.*

Let a = the base AB,

  b = the difference between AC and BC,

  x = the hypothenuse AC, and

  y = the perpendicular CB.

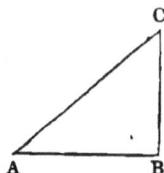

Then, $x - y = b$, by the conditions, $\cdot$ $\cdot$ $\cdot$ $\cdot$ (1)

  $x^2 = a^2 + y^2$ (Bk. IV. Prop. 11), $\cdot$ $\cdot$ $\cdot$ (2)

From equation (1) we have,

$$x = y + b;$$

by squaring, $x^2 = y^2 + 2by + b^2.$

Substituting this value of x^2 in equation (2),

$$y^2 + 2by + b^2 = a^2 + y^2;$$

hence, $2by = a^2 - b^2$,   or, $y = \dfrac{a^2 - b^2}{2b};$

substituting this value in equation (1), we readily find

$$x = \frac{a^2 + b^2}{2b}.$$

PROP. LXXIII.—*In a right-angled triangle, having given the hypothenuse and the difference between the base and perpendicular, to determine the triangle.*

Let a = the hypothenuse AC,

  b = difference between AB and BC,

$x =$ the base AB,   and

$y =$ the perpendicular BC.

Then, $x - y = b$, by the conditions, $\cdot$ $\cdot$ (1)

$a^2 = x^2 + y^2$, (Bk. IV, Prop. 11) $\cdot$ (2)

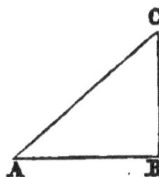

From equation (1), we have,

$$x = y + b;$$

and by squaring,

$$x^2 = y^2 + 2by + b^2.$$

Substituting this value of x^2 in equation (2), we have,

$$a^2 = y^2 + 2by + b^2 + y^2;$$

and by transposing and reducing,

$$y^2 + by = \frac{a^2 - b^2}{2};$$

hence, $y = \dfrac{-b + \sqrt{2a^2 - b^2}}{2};$ $y = -\left(\dfrac{b + \sqrt{2a^2 - b^2}}{2}\right);$

Substituting either of these values in equation (1), we readily find the corresponding value of x.

NOTE.—The *positive value* of the unknown quantity *generally* fulfils the conditions of the problem, understood in its arithmetical sense.

The negative value will always satisfy the *conditions of the equation:* with its sign changed, it may be regarded as the answer to a problem which differs from the one proposed only in this: *that certain quantities which were additive have become subtractive, and the reverse.*

PROP. LXXIV.—*Having given the area of a rectangle inscribed in a given triangle, to determine the sides of the rectangle.*

By a given triangle, we mean one whose sides are all known, or given.

1st. *To find the perpendicular and segments of the base:*

Let ABC be a triangle, in which the three
sides are given : viz.,

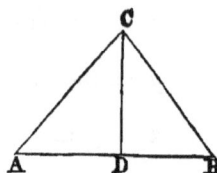

$$b = AB, \quad a = AC, \quad \text{and} \quad c = BC.$$

Let CD be drawn perpendicular to AB, and

let $y = CD$, and $x = AD$; then will $b - x = DB$.

Then, $$a^2 = x^2 + y^2 \quad \cdots \quad \cdots \quad (1)$$
$$c^2 = y^2 + x^2 - 2bx + b^2 \quad \cdots \quad (2) ;$$

substituting in equation (2) the value of $x^2 + y^2 = a^2$, from
equation (1),

$$c^2 = a^2 - 2bx + b^2 ; \quad \text{hence,}$$

$$x = \frac{a^2 + b^2 - c^2}{2b} ;$$

and substituting this value in equation (1), we find the altitude of
the triangle.

2d. *To find the sides of the rectangle.*

Suppose the rectangle to be inscribed in the
triangle ACB.

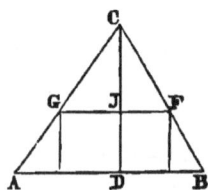

Let $d =$ the area of the rectangle,

　$x =$ its base,

　$y =$ its altitude, and

　$h = CD$, the altitude of the triangle.

Then, by similar triangles,

$$AB \quad : \quad CD \quad : : \quad GF \quad : \quad CJ ; \quad \text{that is,}$$
$$b \quad : \quad h \quad : : \quad x \quad : \quad h - y ; \quad \text{hence,}$$
$$hx = bh - by \quad \cdots \quad (1),$$

and $$xy = d, \quad \text{by the conditions, (2).}$$

From equation (2), we have $$x = \frac{d}{y} ;$$

substituting this value in equation (1), we have,

$$h\frac{d}{y} = bh - by; \quad \text{clearing of fractions,}$$

$$hd = bhy - by^2; \quad \text{or,}$$

$$y^2 - hy = \frac{-hd}{b}; \quad \text{whence,}$$

$$y = +\frac{h}{2} + \frac{1}{2}\sqrt{\frac{b\,h^2 - 4hd}{b}}; \quad y = +\frac{h}{2} - \frac{1}{2}\sqrt{\frac{b\,h^2 - 4hd}{b}}.$$

Substituting these values of y in equation (1), we find the corresponding values of x.

PROP. LXXV.—*In a triangle, having given the ratio of the two sides, together with both the segments of the base made by a perpendicular from the vertical angle, to determine the triangle.*

Let ACB be a triangle, and CD a line drawn perpendicular to the base AB.

Let

$$AC = x$$
$$CB = y$$
$$CD = z$$
$$AD = a$$
$$DB = b$$
$$c = \text{ratio.}$$

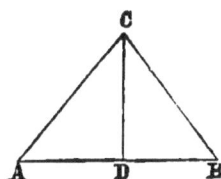

Then, $\dfrac{x}{y} = c$, by the conditions $\cdot$ $\cdot$ $\cdot$ (1)

$$x^2 - a^2 = z^2 \ (\text{Bk. IV., Prop. 11, Cor. 1}) \ \cdot \ \cdot \ (2)$$

$$y^2 - b^2 = z^2 \ (\text{Bk. IV., Prop. 11, Cor. 1}) \ \cdot \ \cdot \ (3).$$

Subtracting (3) from (2), member from member,

$$x^2 - y^2 + b^2 - a^2 = 0; \quad \text{or,} \quad x^2 = y^2 + a^2 - b^2 \ \cdot \ (4).$$

From equation (1) we have

$$x = cy; \quad\quad \text{hence,} \quad\quad x^2 = c^2 y^2 \quad \cdot \quad \cdot \quad \cdot \quad (5);$$

combining (4) and (5), we have

$$c^2 y^2 = y^2 + a^2 - b^2, \quad \text{and} \quad (c^2 - 1) y^2 = a^2 - b^2;$$

hence, $\quad\quad y^2 = \dfrac{a^2 - b^2}{c^2 - 1}; \quad$ and $\quad y = \pm \sqrt{\dfrac{a^2 - b^2}{c^2 - 1}};$

substituting these values in equation (1), we find the corresponding values of x.

PROP. LXXVI.—*In a triangle, having given the base, the sum of the other two sides and the length of a line drawn from the vertical angle to the middle of the base, to find the sides of the triangle.*

Let ABC be a triangle, and CD the line drawn from the vertex C to D, the middle point of the base.

Let
$$AC = x$$
$$BC = y$$
$$AD = b$$
$$CD = a$$
$$AC + BC = s.$$

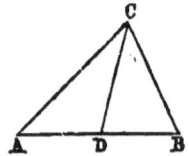

Then, $\quad\quad\quad\quad x + y = s, \quad$ by the conditions $\quad \cdot \quad \cdot \quad (1)$

and $\quad x^2 + y^2 = 2b^2 + 2a^2$, (Bk. IV., Prop. 14) $\quad \cdot \quad \cdot \quad (2);$

equation (1), by transposing and squaring, gives

$$y^2 = x^2 - 2sx + s^2;$$

substituting this value in equation (2), we have

$$2x^2 - 2sx + s^2 = 2b^2 + 2a^2;$$

transposing and reducing, we have

$$x^2 - sx = \frac{2b^2 + 2a^2 - s^2}{2} : \quad \text{whence,}$$

$$x = \frac{s + \sqrt{4b^2 + 4a^2 - s^2}}{2}, \quad \text{and} \quad x = \frac{s - \sqrt{4b^2 + 4a^2 - s^2}}{2}.$$

PROP. LXXVII.—*In a triangle, having given the two sides about the vertical angle, together with a line bisecting that angle and terminating in the base, to find the base.*

Let ACB be a triangle, and CD a line bisecting the angle ACB, and terminating in the base at D.

Let

$$\begin{aligned} AC &= a \\ BC &= b \\ CD &= c \\ AD &= x \\ DB &= y. \end{aligned}$$

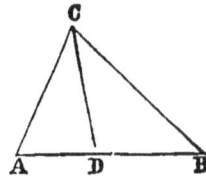

Then, $x : y :: a : b$ (Bk. IV., Prop. 17) ;

whence, $bx = ay \cdots (1)$;

also, $a \times b = c^2 + xy$ (Bk. IV., Prop. 31) $\cdots (2)$.

Multiplying the first equation by x, and the second by a, and subtracting, member from member, we have

$$bx^2 - a^2b = -ac^2 ; \qquad \text{whence}$$

$$x = = \sqrt{\frac{a(ab - c^2)}{b}} ; \qquad y = = \sqrt{\frac{b(ab - c^2)}{a}}.$$

PROP. LXXVIII.—*To determine a right-angled triangle, having given the lengths of two lines drawn from the vertices of the acute angles to the middle points of the opposite sides.*

Let ABC be a right angled triangle, and AD, CE, two lines drawn to the middle points D and E of the opposite sides.

$$\text{Let } AD = a$$
$$CE = b$$
$$AB = 2x$$
$$BC = 2y$$

Then, $4x^2 + y^2 = a^2$ (Bk. IV. Prop. 11)   .   **(1)**

and, $4y^2 + x^2 = b^2$ "   "   "   "   . . **(2)**.

Multiplying equation (1) by 4, and subtracting (2) from it, we have,

$$15x^2 = 4a^2 - b^2; \quad \text{when, } x = \pm \sqrt{\frac{4a^2 - b^2}{15}};$$

and substituting in equation (2) $\quad y = \pm \sqrt{\frac{4b^2 - a^2}{15}}.$

PROP. LXXIX.—*To determine a right-angled triangle, having given the perimeter and the radius of the inscribed circle.*

1st. Let ABC be a right angled triangle, O the centre, and r the radius of the inscribed circle.

Let $p =$ the perimeter; let x denote the length of the equal tangents drawn from A (Bk. III. Prop. 14, sch.); v the length of the equal tangents drawn from B; and y the length of the equal tangents drawn from C.

Then, $AC = x + y$, $AB = x + v$,   and $BC = y + v$;   then,

$$AC + AB + BC = 2x + 2y + 2v = p;$$

transposing and reducing,

$$x + y = \frac{p - 2v}{2} = a, \text{ a known quantity} \quad \cdot \ \cdot \ \cdot \quad \textbf{(1)}.$$

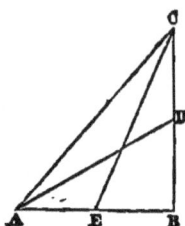

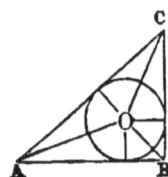

Then, $AB^2 + BC^2 = AC^2$; that is,

$$x^2 + 2vx + v^2 + y^2 + 2vy + v^2 = a^2 \quad \cdot \quad \cdot \quad \cdot \quad (2).$$

2nd. Observe that the double area of each of the triangles AOB, BOC, and AOC, is equal to its base multiplied by the radius of the inscribed circle; and hence, the sum of these products is equal to the sum of the bases multiplied by r; that is, $= r \times p$, a known quantity.

But the base $AB \times BC$ is also equal to double the area of the triangle ABC; hence,

$$(x + v) \times (y + v) = r \times p; \text{ that is,}$$

$$xy + vx + vy + v^2 = r \times p \quad \cdot \quad \cdot \quad \cdot \quad (3);$$

Multiplying both numbers of equation (3) by 2, we have,

$$2xy + 2vx + 2vy + 2v^2 = 2rp \quad \cdot \quad \cdot \quad \cdot \quad (4);$$

subtracting equation (4) from (2), we have,

$$x^2 - 2xy + y^2 = a^2 - 2rp; \text{ whence, by extracting square root,}$$

$$x - y = \pm \sqrt{a^2 - 2rp} = b \quad \cdot \quad \cdot \quad \cdot \quad \cdot \quad \cdot \quad (5);$$

combining (1) and (5) we readily find the values of x and y,

$$x = \frac{a + b}{2}, \text{ and } y = \frac{a - b}{2}.$$

Prop. LXXX.—*To determine a triangle, having given the base, the perpendicular, and the ratio of the two sides.*

Let ACB be a triangle, and CD perpendicular to the base AB.

Let $\qquad$ $AC = y$,

$\qquad$ $r = $ ratio;

then, $\qquad$ $ry = CB$

$\qquad$ $AB = b$

$\qquad$ $DB = b - x$

$\qquad$ $AD = x$

$\qquad$ $CD = h$.

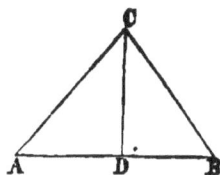

Then, $\quad y^2 = x^2 + h^2$; and $\quad r^2 y^2 = b^2 - 2bx + x^2 + h^2$.

Multiplying the first equation by r^2, and subtracting,

$$0 = b^2 - 2bx + x^2 + h^2 - r^2x^2 - r^2h^2; \quad \text{whence,}$$

$$(1 - r^2) x^2 - 2bx = (r^2 - 1) h^2 - b^2; \quad \text{or,}$$

$$x^2 - \frac{2b}{1 - r^2} \cdot x = \frac{(r^2 - 1) h^2 - b^2}{1 - r^2}; \quad \text{whence,}$$

$$x = \frac{b \pm \sqrt{[(r^2 - 1) h^2 - b^2] (1 - r^2) + b^2}}{1 - r^2}.$$

Prop. LXXXI.—*To determine a right-angled triangle, having given the hypothenuse, and the side of the inscribed square.*

Let ACB be a right-angled triangle, and FDEB an inscribed square.

Let $\quad$ AC $= h$, and $\quad$ DF $= s$: also, denote AB by x, and BC by y; then CE$=y-s$.

Then, $\quad$ AB $\;:\;$ BC $\;::\;$ DE $\;:\;$ EC;

that is, $\quad x \;:\; y \;::\; s \;:\; y - s$;

whence, $\quad xy - sx = sy$; or,

$$xy = sy + sx = s(x + y) \quad \cdot \quad \cdot \quad (1);$$

also, $\quad x^2 + y^2 = h^2$ (Bk. IV., Prop. 11) $\cdot \; \cdot \;$ (2).

If to equation 2, we add twice equation (1), we have

$$x^2 + 2xy + y^2 = 2s(x + y) + h^2; \quad \text{or,}$$

$$(x + y)^2 - 2s(x + y) = h^2 \quad \cdot \quad \cdot \quad (3),$$

which is an equation of the second degree, in which the unknown quantity is $x + y$; hence,

$$x + y = s + \sqrt{h^2 + s^2} \quad \cdot \quad \cdot \quad (4).$$

Combining equations (1) and (4), we have

$$xy = s^2 + s\sqrt{h^2 + s^2} \quad \cdot \quad \cdot \quad (5).$$

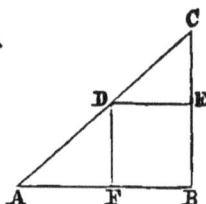

From the square of equation **(4)**, subtract 4 times equation **(5)**, and we have

$$x^2 - 2xy + y^2 = h^2 - 2s^2 - 2s\sqrt{h^2 + s^2} \quad \cdot \ \cdot \quad (6) \ ;$$

extracting the square root of both members,

$$x - y = \sqrt{h^2 - 2s^2 - 2s\sqrt{h^2 + s^2}} \quad \cdot \ \cdot \quad (7) \ ,$$

combining equations **(4)** and **(7)**, we have

$$x = \frac{s + \sqrt{h^2 + s^2} + \sqrt{h^2 - 2s^2 - 2s\sqrt{h^2 + s^2}}}{2} \quad \cdot \ \cdot \quad (8),$$

$$y = \frac{s + \sqrt{h^2 + s^2} - \sqrt{h^2 - 2s^2 - 2s\sqrt{h^2 + s^2}}}{2} \quad \cdot \ \cdot \quad (9).$$

PROP. LXXXII.—*To determine the radii of three equal circles, described within and tangent to, a given circle, and also tangent to each other.*

Let O be the centre of the given circle, and A, B and C, the centres of the equal inscribed circles.

Denote the radius of the given circle by R, and the equal radii of the inscribed circles by r.

Joining the centres A and C, C and B, B and A, by straight lines, we have the equilateral triangle ABC, each of whose sides is $2r$. Draw COD and prolong it to E, and it will be perpendicular to AB.

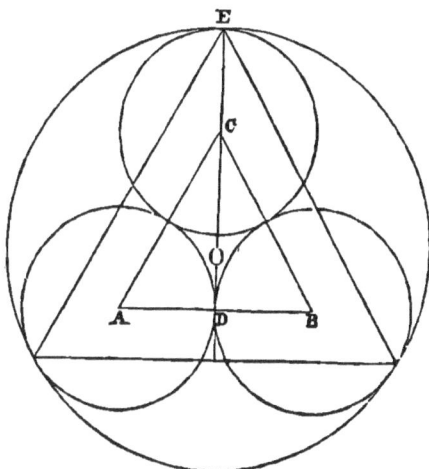

Then, in the right-angled triangle ACD, we have

$$CD^2 = AC^2 - AD^2, \quad \text{or}$$

$$CD^2 = 4r^2 - r^2 = 3r^2, \quad \text{or} \quad CD = r\sqrt{3}.$$

But since O is the point at which lines drawn from the vertices of the angles to the middle points of the opposite sides, in both triangles, intersect each other, it follows, from Cor. of Prob. 21, that

$$CO = \tfrac{2}{3} CD = \tfrac{2}{3} r \sqrt{3}; \quad \text{hence,}$$

$$OE = R = r + \tfrac{2}{3} r \sqrt{3}; \quad \text{hence, finally,}$$

$$r = \frac{3\,R}{3 + 2\sqrt{3}} = \frac{R}{1 + 2\sqrt{\tfrac{1}{3}}}.$$

PROP. LXXXIII.—*In a right-angled triangle, having given the perimeter and the perpendicular let fall from the right angle on the hypothenuse, to determine the triangle.*

Let ACB be a right-angled triangle, right angled at C; and let CD be drawn perpendicular to the hypothenuse AB.

Let $p =$ the perimeter, and

             $h =$ the perpendicular CD.

Denote AC by x, and CB by y;

then, $AB = \sqrt{x^2 + y^2}$,   and

$$AC + CB + AB = p; \quad \text{that is,}$$

$$x + y + \sqrt{x^2 + y^2} = p \quad \cdot \quad (1).$$

Again, $xy =$ double the area of ACB $\cdot \cdot$ (2),

and $h\sqrt{x^2 + y^2} =$ double area $\cdot \cdot$ (3);

hence, $xy = h\sqrt{x^2 + y^2} \quad \cdot \cdot \cdot \cdot$ (4)

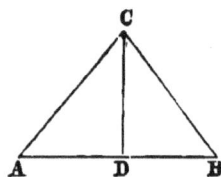

transposing in equation (1) and squaring,

$$(x + y)^2 = (p - \sqrt{x^2 + y^2})^2,$$

$$x^2 + 2xy + y^2 = p^2 - 2p\sqrt{x^2 + y^2} + x^2 + y^2,$$

or $$2xy = p^2 - 2p\sqrt{x^2 + y^2} \quad \cdot \quad \cdot \quad (5);$$

multiplying both members of equation (4) by 2,

$$2xy = 2h\sqrt{x^2 + y^2} \quad \cdot \quad \cdot \quad \cdot \quad (6);$$

combining equations (5) and (6),

$$2h\sqrt{x^2 + y^2} = p^2 - 2p\sqrt{x^2 + y^2}; \quad \text{hence,}$$

$$\sqrt{x^2 + y^2} = \frac{p^2}{2(h + p)} \quad \cdot \quad \cdot \quad \cdot \quad (7);$$

substituting this value of $\sqrt{x^2 + y^2}$, in equation (4),

$$xy = \frac{hp^2}{2(h + p)} \quad \cdot \quad \cdot \quad \cdot \quad (8):$$

squaring both members of equation (7), we have

$$x^2 + y^2 = \frac{p^4}{4(h + p)^2};$$

adding and subtracting 2 times each member of (8),

$$x^2 + 2xy + y^2 = \frac{p^2(p^2 + 4ph + 4h^2)}{4(h + p)^2} \quad \cdot \quad \cdot \quad \cdot \quad (9),$$

$$x^2 - 2xy + y^2 = \frac{p^2(p^2 - 4ph - 4h^2)}{4(h + p)^2} \quad \cdot \quad \cdot \quad (10).$$

Extracting the square roots of equations (9) and (10),

$$x + y = \frac{p(p + 2h)}{2(h + p)} \quad \cdot \quad \cdot \quad \cdot \quad \cdot \quad (11),$$

$$x - y = \frac{p\sqrt{p^2 - 4ph - 4h^2}}{2(h + p)} \quad \cdot \quad (12);$$

hence, $\quad x = \dfrac{p\,(p+2h) + p\,\sqrt{p^2 - 4ph - 4h^2}}{4\,(h+p)}$ $\quad \cdot \quad$ (13),

$y = \dfrac{p\,(p+2h) - p\,\sqrt{p^2 - 4ph - 4h^2}}{4\,(h+p)}$ $\quad \cdot \quad$ (14).

PROP. LXXXIV.—*To determine a right-angled triangle, having given the hypothenuse and the difference of two lines drawn from the two acute angles to the centre of the inscribed circle.*

Let ABC be a right-angled triangle, right-angled at B, and AO, OC, two lines drawn from the acute angles to the centre O of the inscribed circle.

Let $AC = h$; $AO - OC = d$, $OC = x$;
then, $\qquad AO = x + d$

Produce AO, and from C draw CD perpendicular to the prolongation, meeting it at D. Then, since the sum of the angles BAC and ACB is equal to a right-angle (Bk. I, Prop. 25, Cor. 4), and since the lines AO and CO bisect these angles (Bk. III, Prob. 14), OAC + ACO is equal to a half a right-angle.

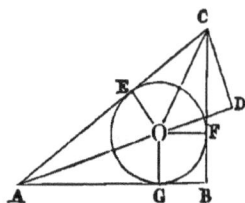

Since the outward angle COD is equal to the sum of the inward angles (Bk. I, Prop. 25, Cor. 6), it is equal to half a right-angle; and hence, OCD is equal to half a right-angle, and hence OD and CD are equal. Denote either by z; then,

$$x^2 = z^2 + z^2; \quad \text{and} \quad z = x\,\sqrt{\tfrac{1}{2}}.$$

Then, $\qquad AC^2 = AD^2 + CD^2;$ that is,

$$h^2 = (x + d + x\,\sqrt{\tfrac{1}{2}})^2 + (x\,\sqrt{\tfrac{1}{2}})^2;$$

and by reduction, $\quad (2 + \sqrt{2})\,x^2 + d\,(2 + \sqrt{2})\,x = h^2 - d^2;$ or,

$$x^2 + dx = \frac{h^2 - d^2}{2 + \sqrt{2}}$$

hence, $\quad x = -\frac{1}{2}d + \frac{1}{2}\sqrt{\dfrac{4h^2 - (2-\sqrt{2})\,d^2}{2+\sqrt{2}}},$

$$x = -\frac{1}{2}d - \frac{1}{2}\sqrt{\dfrac{4h^2 - (2-\sqrt{2})\,d^2}{2+\sqrt{2}}}.$$

Let $\quad OC = x = m, \quad AO = m + d, \quad CD = z = m\sqrt{\frac{1}{2}},$

$$AB = y, \quad \text{and} \quad CB = u.$$

Then, since the triangles ACD and AOG are similar,

$$h \;:\; m\sqrt{\tfrac{1}{2}} \;::\; m + d \;:\; OG = r = \frac{m\sqrt{\frac{1}{2}} \times (m+d)}{h}; \text{ and}$$

$$h \;:\; m + d + m\sqrt{\tfrac{1}{2}} \;::\; m + d \;:\; AG = \frac{(m+d)\,(m+d+m\sqrt{\frac{1}{2}})}{h}$$

$$AB = y = AG + GB = \frac{(m+d)\,(m+d+m\sqrt{\frac{1}{2}}) + m\sqrt{\frac{1}{2}}(m+d)}{h}$$

$$= \frac{(m+d)\,(m+d+m\sqrt{2})}{h} = b; \quad \text{and}$$

$$u = \sqrt{h^2 - b^2}.$$

Prop. LXXXV.—*To determine a triangle, having given the base, the perpendicular, and the difference of the two other sides.*

Let ACB be a triangle, and CD perpendicular to the base AB.

Let $\quad AB = b, \quad CD = a, \quad AC - BC = d,$

$\quad\quad BC = x, \quad AC = x+d, \quad \text{and} \quad DB = z,$

and $\quad\quad AD = b - z.$

Then, $\quad\quad AC = \sqrt{a^2 + (b-z)^2};$

and $\quad\quad CB = \sqrt{a^2 + z^2}; \quad$ hence,

$$\sqrt{a^2 + (b-z)^2} - \sqrt{a^2 + z^2} = d.$$

Transposing and squaring, we have

$$a^2 + (b-z)^2 = d^2 + 2d\sqrt{a^2 + z^2} + a^2 + z^2;$$

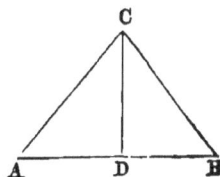

squaring $b - z$, in the first member, we have

$$a^2 + b^2 - 2bz + z^2 = d^2 + 2d \sqrt{a^2 + z^2} + a^2 + z^2;$$

whence, by reducing,

$$(b^2 - d^2) - 2bz = 2d \sqrt{a^2 + z^2};$$

squaring both members, we have

$$(b^2 - d^2)^2 - 4b(b^2 - d^2) \cdot z + 4b^2 z^2 = 4a^2 d^2 + 4d^2 z^2;$$

transposing, and collecting the terms,

$$4(b^2 - d^2) \cdot z^2 - 4b(b^2 - d^2) \cdot z = 4a^2 d^2 - (b^2 - d^2)^2; \quad \textbf{or,}$$

$$z^2 - bz = \frac{4a^2 d^2 - (b^2 - d^2)^2}{4(b^2 - d^2)}; \quad \text{whence,}$$

$$z = \frac{b}{2} \pm \frac{1}{2} \sqrt{\frac{4a^2 d^2 + 2b^2 d^2 - d^4}{b^2 - d^2}};$$

from which the sides AC and CB are easily found.

PROP. LXXXVI.—*To determine a triangle, having given the base, the perpendicular, and the rectangle of the other sides.*

Let ACB be a triangle, and CD a line drawn perpendicular to the base AB.

Let   AB $= b$,   CD $= d$,   AC $\times$ CB $= q$;

AC $= x$,   and   BC $= y$;   then,

$$xy = q \quad \cdot \quad \cdot \quad \cdot \quad (1).$$

AD $= \sqrt{x^2 - d^2}$,   and   DB $= \sqrt{y^2 - d^2}$;   hence,

$$\sqrt{x^2 - d^2} + \sqrt{y^2 - d^2} = b \quad \cdot \quad \cdot \quad \cdot \quad (2).$$

Transposing, and then squaring both members of equation (2), we have

$$x^2 - d^2 = b^2 - 2b\sqrt{y^2 - d^2} + y^2 - d^2;$$

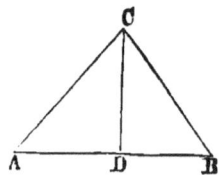

whence, $(x^2 - y^2) - b^2 = -2b\sqrt{y^2 - d^2}$; squaring again,

$x^4 - 2x^2y^2 + y^4 - 2b^2(x^2 - y^2) + b^4 = 4b^2y^2 - 4b^2d^2$; reducing,

$x^4 - 2x^2y^2 + y^4 - 2b^2(x^2 + y^2) = -b^4 - 4b^2d^2 \quad \cdot \quad \cdot \quad (3).$

Adding 4 times the square of each member of (1),

$x^4 + 2x^2y^2 + y^4 - 2b^2(x^2 + y^2) = 4q^2 - b^4 - 4b^2d^2$; or,

$(x^2 + y^2)^2 - 2b^2(x^2 + y^2) = 4q^2 - b^4 - 4b^2d^2.$

Regarding $x^2 + y^2$ as a single unknown quantity, we have

$$x^2 + y^2 = b^2 \pm \sqrt{4q^2 - 4b^2d^2} \quad \cdot \quad \cdot \quad (4);$$

then, by adding twice each member of equation (1),

$x^2 + 2xy + x^2 = 2q + b^2 \pm \sqrt{4q^2 - 4b^2d^2}$; or,

$$x + y = \sqrt{2q + b^2 \pm \sqrt{4q^2 - 4b^2d^2}} = m;$$

subtracting from equation (4), twice equation (1),

$x^2 - 2xy + x^2 = -2q + b^2 \pm \sqrt{4q^2 - 4b^2d^2}$; or,

$$x - y = \pm\sqrt{-2q + b^2 \pm \sqrt{4q^2 - 4b^2d^2}} = n;$$

hence, $\quad x = \dfrac{m+n}{2}, \quad$ and $\quad y = \dfrac{m-n}{2}.$

PROP. LXXXVII.—*To determine a triangle, having given the lengths of three lines drawn from the three angles to the middle of the opposite sides.*

Let ACB be a triangle, and AE, BG, CD, three lines drawn from the vertices to the middle points of the opposite sides.

Let CD=a, AE=b, BG=c, AC=x,
BC=y, and AB=z: then,

$x^2 + y^2 = 2a^2 + \dfrac{z^2}{2}$ (Bk. IV. Prop. 14),

$x^2 + z^2 = 2b^2 + \dfrac{y^2}{2}$; $y^2 + z^2 = 2c^2 + \dfrac{x^2}{2}$;

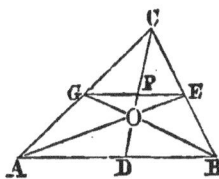

clearing of fractions and transposing, we have

$$2x^2 + 2y^2 - z^2 = 4a^2 \quad \cdots \quad (1)$$

$$2x^2 + 2z^2 - y^2 = 4b^2 \quad \cdots \quad (2)$$

$$2y^2 + 2z^2 - x^2 = 4c^2 \quad \cdots \quad (3);$$

subtracting equation (2) from (1), we have

$$3y^2 - 3z^2 = 4\,(a^2 - b^2) \quad \cdot \quad \cdot \quad (4);$$

multiplying equation (3) by 2, and adding to (2),

$$3y^2 + 6z^2 = 8c^2 + 4b^2 \quad \cdot \quad \cdot \quad (5);$$

subtracting equation (4) from (5),

$$9z^2 = 8c^2 + 4b^2 - 4a^2 + 4b^2; \quad \text{hence,}$$

$$z = \pm \tfrac{2}{3}\sqrt{2b^2 + 2c^2 - a^2}$$

$$x = \pm \tfrac{2}{3}\sqrt{2a^2 + 2b^2 - c^2}$$

$$y = \pm \tfrac{2}{3}\sqrt{2a^2 + 2c^2 - b^2}.$$

Cor. Through G and E, the middle points of AC and BC, draw GE; then will GE be parallel to the base AB (Bk. IV. Prop. 16) : and since AC is equal to twice CG, AD will be equal to twice GP, or its equal PE.

But the triangles AOD and POE are similar; and, since AD is equal to twice PE, AO is equal to twice OE; that is,

If three lines be drawn from the vertices of the three angles of a triangle to the middle points of the opposite sides, the distance from either vertex to the point of intersection, will be two-thirds of the bisecting line.

Prop. LXXXVIII.—*In a triangle, having given the three sides, to find the radius of the inscribed circle.*

Let ABC be a triangle, CD perpendicular to the base, O the centre, and OP the radius of the inscribed circle.

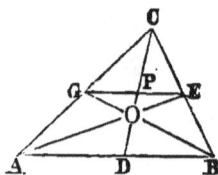

Let $BC = a$, $AB = c$, $AC = b$,

$OP = r$, $AD = z$, $CD = x$.

$$x^2 = b^2 - z^2 \quad . \quad . \quad . \quad (1),$$

$$x^2 = a^2 - (c - z)^2; \text{ whence,}$$

$$b^2 - z^2 = a^2 - c^2 + 2cz - z^2; \text{ therefore,}$$

$$z = \frac{b^2 + c^2 - a^2}{2c};$$

combining with equation (1),

$$x = \frac{\sqrt{4b^2c^2 - (b^2 + c^2 - a^2)^2}}{2c};$$

area $= (a + b + c) \tfrac{1}{2} r = \tfrac{1}{2} cx$; whence,

$$r = \frac{cx}{a + b + c} = \frac{\sqrt{4b^2c^2 - (b^2 + c^2 - a^2)^2}}{2 (a + b + c)}.$$

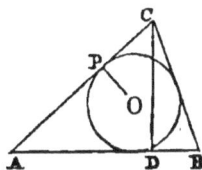

PROP. LXXXIX.—*To determine a right-angled triangle, having given the side of the inscribed square, and the radius of the inscribed circle.*

Let ABC be a right-angled triangle, with a square and circle both inscribed.

Let $AB = x$, $BC = y$, and $AC = z$: denote the side of the square by s, and the radius of the circle by r.

Then, $x + y - z = 2r$ (Prop. LXXIX, Key). (1),

$$x^2 + y^2 = z^2 \quad . \quad . \quad . \quad . \quad . \quad . \quad . \quad . \quad . (2),$$

and $x + y = s + \sqrt{z^2 + s^2}$ (Prop. LXXXI, Key) . . (3).

Combining equations (1) and (3), and transposing,

$$z + 2r - s = \sqrt{z^2 + s^2}; \text{ squaring,}$$

$$z^2 + 4r^2 + s^2 + 4rz - 2sz - 4rs = z^2 + s^2;$$

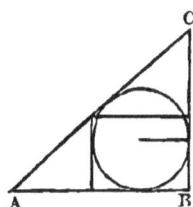

whence, $z = \dfrac{2rs - 2r^2}{2r - s}$ . . . (4);

combining equations (4) and (1),

$$x + y = \frac{2r^2}{2r - s} = m, \text{ and equation (2) gives}$$

$$x^2 + y^2 = \frac{(2rs - 2r^2)^2}{(2r - s)^2} = n^2; \text{ whence,}$$

$$x = \tfrac{1}{2}\left(m + \sqrt{2n^2 - m^2}\right),$$

$$y = \tfrac{1}{2}\left(m - \sqrt{2n^2 - m^2}\right).$$

Prop. XC.—*To determine a right-angled triangle, having given the hypothenuse and radius of the inscribed circle.*

Let ABC be a right-angled triangle, and O the centre of the inscribed circle.

Let   AC = h,   AB = x,   BC = y,

and r = the radius of the circle.

$x + y = h + 2r$ (Prop. LXXIX, Key) . (1)

$x^2 + y^2 = h^2$ . . . . (2).

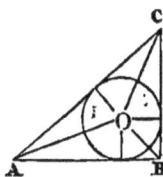

Since the perimeter is equal to $2h + 2r$, and since four times the area is equal to the perimeter into $2r$,

$$2xy = 4r^2 + 4hr \quad . . . \quad (3);$$

subtracting equation (3) from (2), and extracting the square root,

$$x - y = \sqrt{h^2 - 4r^2 - 4rh} \quad . . . \quad (4);$$

combining (4) and (1),

$$x = \frac{h + 2r + \sqrt{h^2 - 4r^2 - 4rh}}{2},$$

$$y = \frac{h + 2r - \sqrt{h^2 - 4r^2 - 4rh}}{2}.$$

THE NATIONAL SERIES OF READERS.

COMPLETE IN TWO INDEPENDENT PARTS.

I.
THE NATIONAL READERS.
By PARKER & WATSON.

No. 1.—National Primer, ...	*64 pp., 16mo,*	\$0 25
No. 2.—National First Reader, . . .	*128 pp., 16mo,*	38
No. 3.—National Second Reader, . .	*224 pp., 16mo,*	63
No. 4.—National Third Reader, . .	*288 pp., 12mo,*	1 00
No. 5.—National Fourth Reader, . .	*432 pp., 12mo,*	1 50
No. 6.—National Fifth Reader, . .	*600 pp., 12mo,*	1 88

National Elementary Speller, . . .	*160 pp., 16mo,*	25
National Pronouncing Speller, . . .	*188 pp., 12mo,*	50

II.
THE INDEPENDENT READERS.
By J. MADISON WATSON.

The Independent First Reader, . .	*80 pp., 16mo,*	25
The Independent Second Reader, .	*160 pp., 16mo,*	50
The Independent Third Reader, . .	*240 pp., 16mo,*	75
The Independent Fourth Reader, . .	*264 pp., 12mo,*	1 00
The Independent Fifth Reader, . .	*336 pp., 12mo,*	1 25
The Independent Sixth Reader, . .	*474 pp., 12mo,*	1 50

The Independent Child's Speller (Script),	*80 pp., 16mo,*	25
The Independent Youth's Speller (Script),	*168 pp., 12mo,*	50
The Independent Spelling Book, . .	*160 pp., 16mo,*	25

₃ **The Readers** constitute two complete and entirely distinct series, either of which is adequate to every want of the best schools. The Spellers may accompany either Series.

PARKER & WATSON'S NATIONAL READERS,

The salient features of these works which have combined to render them so popular may be briefly recapitulated as follows :

1. **THE WORD-BUILDING SYSTEM.**—This famous progressive method for young children originated and was copyrighted with these books. It constitutes a process with which the beginner with *words* of one letter is gradually introduced to additional lists formed by prefixing or affixing single letters, and is thus led almost insensibly to the mastery of the more difficult constructions. This is one of the most striking modern improvements in methods of teaching.

2. **TREATMENT OF PRONUNCIATION.**—The wants of the youngest scholars in this department are not overlooked. It may be said that from the first lesson the student by this method need never be at a loss for a prompt and accurate rendering of every word encountered.

3. **ARTICULATION AND ORTHOEPY** are considered of primary importance.

4. **PUNCTUATION** is inculcated by a series of interesting *reading lessons*, the simple perusal of which suffices to fix its principles indelibly upon the mind.

5. **ELOCUTION.** Each of the higher Readers (3d, 4th and 5th) contains elaborate, scholarly, and thoroughly practical treatises on elocution. This feature alone has secured for the series many of its warmest friends.

6. **THE SELECTIONS** are the crowning glory of the series. Without exception it may be said that no volumes of the same size and character contain a collection so diversified, judicious, and artistic as this. It embraces the choicest gems of English literature, so arranged as to afford the reader ample exercise in every department of style. So acceptable has the taste of the authors in this department proved, not only to the educational public but to the reading community at large, that thousands of copies of the Fourth and Fifth Readers have found their way into public and private libraries throughout the country, where they are in constant use as manuals of literature, for reference as well as perusal.

7. **ARRANGEMENT.** The exercises are so arranged as to present constantly alternating practice in the different styles of composition, while observing a definite plan of progression or gradation throughout the whole. In the higher books the articles are placed in formal sections and classified topically, thus concentrating the interest and inculcating a principle of association likely to prove valuable in subsequent general reading.

8. **NOTES AND BIOGRAPHICAL SKETCHES,** These are full and adequate to every want. The biographical sketches present in pleasing style the history of every author laid under contribution.

9. **ILLUSTRATIONS,** These are plentiful, almost profuse, and of the highest character of art. They are found in every volume of the series as far as and including the Third Reader.

10. **THE GRADATION** is perfect. Each volume overlaps its companion preceding or following in the series, so that the scholar, in passing from one to another, is only conscious, by the presence of the new book, of the transition.

11. **THE PRICE** is reasonable. The National Readers contain more matter than any other series in the same number of volumes published. Considering their completeness and thoroughness they are much the cheapest in the market.

12. **BINDING.** By the use of a material and process known only to themselves, in common with all the publications of this house, the National Readers are warranted to outlast any with which they may be compared—the ratio of relative durability being in their favor as two to one.

WATSON'S INDEPENDENT READERS.

This Series is designed to meet a general demand for smaller and cheaper books than the National Series proper, and to serve as well for intermediate volumes of the National Readers in large graded schools requiring more books than one ordinary series will supply.

Beauty. The most casual observer is at once impressed with the unparalleled mechanical beauty of the Independent Readers. The Publishers believe that the æsthetic tastes of children may receive no small degree of cultivation from their very earliest school books, to say nothing of the importance of making study attractive by all 'such artificial aids that are legitimate. In accordance with this view, not less than $25,000 was expended in their preparation before publishing, with a result which entitles them to be considered "The Perfection of Common School Books."

Selections. They contain, of course, none but entirely new selections. These are arranged according to a strictly progressive and novel method of developing the elementary sounds in order in the lower numbers, and in all, with a view to topics and general literary style. The mind is thus led in fixed channels to proficiency in every branch of good reading, and the evil results of ' scattering ' as practised by most school-book authors, avoided.

The Illustrations, as may be inferred from what has been said, are elegant beyond comparison. They are profuse in every number of the series from the lowest to the highest. This is the only series published of which this is true.

The Type is semi-phonetic, the invention of Prof. Watson. By it every letter having more than one sound is clearly distinguished in all its variations without in any way mutilating or disguising the normal form of the letter.

Elocution is taught by prefatory treatises of constantly advancing grade and completeness in each volume, which are illustrated by wood-cuts in the lower books, and by black-board diagrams in the higher. Prof. Watson is the first to introduce Practical Illustrations and Black-board Diagrams for teaching this branch.

Foot Notes on every page afford all the incidental instruction which the teacher is usually required to impart. Indices of words refer the pupil to the place of their first use and definition. The Biographies of Authors and others are in every sense excellent.

Economy. Although the number of pages in each volume is fixed at the minimum, for the purpose recited above, the utmost amount of matter available without overcrowding is obtained in the space. The pages are much wider and larger than those of any competitor and contain *twenty per cent* more matter than any other series of the same type and number of pages.

All the Great Features. Besides the above all the popular features of the National Readers are retained except the Word-Building system. The latter gives place to an entirely new method of progressive development, based upon some of the best features of the Word System, Phonetics and Object Lessons.

NATIONAL READERS.

ORIGINAL AND "INDEPENDENT" SERIES.

SPECIMEN TESTIMONIALS.

From D. H. Harris, *Supt. Public Schools, Hannibal, Mo.*
The National Series of Readers are now in use in our public schools, and I regard them *the best* that I have ever examined or used.

From Hon. J. K. Jillson, *Supt. of Education, State of South Carolina.*
I have carefully examined your new and beautiful Series of Readers known as "The Independent Readers," and do not hesitate to recommend it as the finest and most excellent ever presented to the public.

From D. N. Rook, *Sec. of School Board, Williamsport, Pa.*
I would say that Parker & Watson's Series of Readers and Spellers give the best satisfaction in our schools of any Series of Readers and Spellers that have ever been used. There is nothing published for which we would exchange them

From Prof. H. Seele, *New Braunfels Academy, Texas.*
I recommend the National Readers for four good reasons : (1.) The printing, en graving, and binding is excellent. (2.) They contain choice selections from English Literature. (3.) They inculcate good morals without any sectarian bias. (4.) They are truly *National*, because they teach pure patriotism and not sectional prejudice.

From S. Findley, *Supt. Akron Schools, Ohio.*
We use no others, and have no desire to. They give entire satisfaction. We like the freshness and excellence of the selections. We like the biographical notes and the definitions at the foot of the page. We also like the white paper and clear and beautiful type. In short, we do not know where to look for books which would be so satisfactory both to teachers and pupils.

From Pres. Robert Allyn, *McKendree College, Ill.*
Since my connection with this college, we have used in our preparatory depart- ment the Series of Readers known as the "National Readers," compiled by Parker & Watson, and published by Messrs. A. S. Barnes & Co. They are *excellent ;* afford choice selections; contain the right system of elocutionary instruction, and are well printed and bound so as to be serviceable as well as interesting. I can com- mend them as among the excellent means used by teachers to make their pupils proficient in that noblest of school arts, Good Reading.

From W. T. Harris, *Supt. Public Schools, St. Louis, Mo.*
I have to admire these excellent selections in prose and verse, and the carefu arrangement which places first what is easy of comprehension, and proceeds gradu ally to what is difficult. I find the lessons so arranged as to bring together differ ent treatments of the same topic, thereby throwing much light on the pupil's path, and I doubt not adding greatly to his progress. The proper variety of subjects chosen, the concise treatise on elocution, the beautiful typography and substantial binding—all these I find still more admirable than in the former series of National Readers, which I considered *models* in these respects.

From H. T. Phillips, Esq., *of the Board of Education, Atlanta, Ga.*
The Board of Education of this city have selected for use in the public school of Atlanta the entire series of your Independent Readers, together with Steele's Chemistry and Philosophy. As a member of the Board, and of the Committee on Text-books, the subject of Readers was referred to me for examination. I gave a pretty thorough examination to ten (10) different series of Readers, and in endea- voring to arrive at a decision upon the sole question of merit, and entirely inde- pendent of any extraneous influence, I very cordially recommended the Independent Series. This verdict was approved by the Committee and adopted by the Board.

From Report of Rev. W. T. Brantly, D.D., *late Professor of Belles Lettres. Univer- sity of Georgia, on "Text-Books in Reading," before the Teachers' Convention of Georgia, May 4, 1870.*

The *National Series*, by Parker & Watson, is deserving of its high reputation. The Primary Books are suited to the weakest capacity ; whilst those more advanced supply instructive illustration on all that is needed to be known in connection with the art.

4

WATSON'S CHILD'S SPELLER.

THE INDEPENDENT CHILD'S SPELLER.

Price 25 Cents.

This unique book, published in 1872, is the first to be consistently printed in imitation of writing; that is, it teaches orthography as we use it. It is for the smallest class of learners, who soon become familiarized with words by their forms, and learn to read writing while they spell.

EXTRACT FROM THE PREFACE.

Success in teaching English orthography is still exceptional, and it must so continue until the principles involved are recognized in practice. Form is foremost: the eye and the hand must be trained to the formation of words; and since spelling is a part of writing, the written form only should be used. The laws of mental association, also—especially those of resemblance, contrast, and contiguity in time and place—should receive such recognition in the construction of the text-book as shall insure, whether consciously or not, their appropriate use and legitimate results. Hence, the spelling-book, properly arranged, is a necessity from the first; and, though primers, readers, and dictionaries may serve as aids, it can have no competent substitute.

Consistently with these views, the words used in the Independent Child's Speller have such original classifications and arrangements in columns—in reference to location, number of letters, vowel sounds, alphabetic equivalents, and consonant terminations—as exhibit most effectively their formation and pronunciation. The vocabulary is strictly confined to the simple and significant monosyllables in common use. He who has mastered these may easily learn how to spell and pronounce words of more than one syllable.

The introduction is an illustrated alphabet in script, containing twenty-six pictures of objects, and their names, commencing both with capitals and small letters. Part First embraces the words of one, two, and three letters; Part Second, the words of four letters; and Part Third, other monosyllables. They are divided into short lists and arranged in columns, the vowels usually in line, so as to exhibit individual characteristics and similarity of formation. The division of words into paragraphs is shown by figures in the columns. Each list is immediately followed by sentences for reading and writing, in which the same words are again presented with irregularities of form and sound. Association is thus employed, memory tested, and definition most satisfactorily taught.

Among the novel and valuable features of the lessons and exercises, probably the most prominent are their adaptedness for young children and their being printed in exact imitation of writing. The author believes that hands large enough to spin a top, drive a hoop, or catch a ball, are not too small to use a crayon, or a slate and pencil; that the child's natural desire to draw and write should not be thwarted, but gratified, encouraged, and wisely directed; and that since the written form is the one actually used in connection with spelling in after-life, the eye and the hand of the child should be trained to that form from the first. He hopes that this little work, designed to precede all other spelling-books and conflict with none, may satisfy the need so universally recognized of a fit introduction to orthography, penmanship, and English composition.

5

The National Readers and Spellers.

THEIR RECORD.

These books have been adopted by the School Boards, or official authority, of the following important States, cities, and towns—in most cases for exclusive use.

The State of Minnesota.

The State of Missouri.

The State of Alabama.

The State of North Carolina.

The State of Louisiana.

The State of Texas.

New York.
New York City.
Brooklyn.
Buffalo.
Albany.
Rochester.
Troy.
Syracuse.
Elmira.
&c., &c.

Pennsylvania.
Reading.
Lancaster.
Erie.
Scranton.
Carlisle.
Carbondale.
Westchester.
Schuylkill Haven.
Williamsport.
Norristown.
Bellefonte.
Wilkesbarre.
&c., &c.

New Jersey.
Newark.
Jersey City.
Paterson.
Trenton.
Camden.
Elizabeth.
New Brunswick.
Phillipsburg.
Orange.
&c., &c.

Delaware.
Wilmington.

D. C.
Washington.

Illinois.
Chicago.
Peoria.
Alton.
Springfield.
Aurora.
Galesburg.
Rockford.
Rock Island.
&c., &c.

Wisconsin.
Milwaukee.
Fond du Lac.
Oshkosh.
Janesville.
Racine.
Watertown.
Sheboygan.
La Crosse.
Waukesha.
Kenosha.
&c., &c.

Michigan.
Grand Rapids.
Kalamazoo.
Adrian.
Jackson.
Monroe.
Lansing.
&c., &c.

Ohio.
Toledo.
Sandusky.
Conneaut.
Chardon.
Hudson.
Canton.
Salem.
&c., &c.

Indiana.
New Albany.
Fort Wayne.
Lafayette.
Madison.
Logansport.
Indianapolis.

Iowa.
Davenport.
Burlington.
Muscatine.
Mount Pleasant.
&c.

Nebraska.
Brownsville.
Lincoln.
&c.

Oregon.
Portland.
Salem.
&c.

Virginia.
Richmond.
Norfolk.
Petersburg.
Lynchburg.
&c.

South Carolina.
Columbia.
Charleston.

Georgia.
Savannah.

Louisiana.
New Orleans.

Tennessee.
Memphis

SCHOOL-ROOM CARDS.

Baade's Reading Case, ...*$10 00

A frame containing movable cards, with arrangement for showing one sentence at a time, capable of 28,000 transpositions.

Eureka Alphabet Tablet *1 50

Presents the alphabet upon the Word Method System, by which the child will learn the alphabet in nine days, and make no small progress in reading and spelling in the same time.

National School Tablets, 10 Nos. *8 00

Embrace reading and conversational exercises, object and moral lessons, form, color, &c. A complete set of these large and elegantly illustrated Cards will embellish the school-room more than any other article of furniture.

READING.

Fowle's Bible Reader $1 00

The narrative portions of the Bible, chronologically and topically arranged, judiciously combined with selections from the Psalms, Proverbs, and other portions which inculcate important moral lessons or the great truths of Christianity. The embarrassment and difficulty of reading the Bible itself, by course, as a class exercise, are obviated, and its use made feasible, by this means.

North Carolina First Reader 40
North Carolina Second Reader 65
North Carolina Third Reader 1 00

Prepared expressly for the schools of this State, by C. H. Wiley, Superintendent of Common Schools, and F. M. Hubbard, Professor of Literature in the State University.

Parker's Rhetorical Reader 1 00

Designed to familiarize Readers with the pauses and other marks in general use, and lead them to the practice of modulation and inflection of the voice.

Introductory Lessons in Reading and Elocution 75

Of similar character to the foregoing, for less advanced classes.

High School Literature 1 50

Admirable selections from a long list of the world's best writers, for exercise in reading, oratory, and composition. Speeches, dialogues, and model letters represent the latter department.

ORTHOGRAPHY.

SMITH'S SERIES

Supplies a speller for every class in graded schools, and comprises the most complete and excellent treatise on English Orthography and its companion branches extant.

1. Smith's Little Speller · · · · · · · .$ 20

First Round in the Ladder of Learning.

2. Smith's Juvenile Definer · · · · · · · 45

Lessons composed of familiar words grouped with reference to similar signification or use, and correctly spelled, accented, and defined.

3. Smith's Grammar-School Speller · · · · 50

Familiar words, grouped with reference to the sameness of sound of syllables differently spelled. Also definitions, complete rules for spelling and formation of derivatives, and exercises in false orthography.

4. Smith's Speller and Definer's Manual · 90

A complete *School Dictionary* containing 14,000 words, with various other useful matter in the way of Rules and Exercises.

5. Smith's Etymology—Small, 75; Complete . 1 25

The first and only Etymology to recognize the *Anglo-Saxon* our *mother tongue;* containing also full lists of derivatives from the Latin, Greek, Gaelic, Swedish, Norman, &c., &c ; being, in fact, a complete etymology of the language for schools.

Sherwood's Writing Speller · · · · · · · 15

Sherwood's Speller and Definer · · · · · 15

Sherwood's Speller and Pronouncer · · · 15

The Writing Speller consists of properly ruled and numbered blanks to receive the words dictated by the teacher, with space for remarks and corrections. The other volumes may be used for the dictation or ordinary class exercises.

Price's English Speller · · · · · · · · · *15

A complete spelling-book for all grades, containing more matter than " Webster," manufactured in superior style, and sold at a lower price—consequently the cheapest speller extant.

Northend's Dictation Exercises · · · · · 63

Embracing valuable information on a thousand topics, communicated in such a manner as at once to relieve the exercise of spelling of its usual tedium, and combine it with instruction of a general character calculated to profit and amuse.

Wright's Analytical Orthography · · · · 25

This standard work is popular, because it teaches the elementary sounds in a plain and philosophical manner, and presents orthography and orthoepy in an easy, uniform system of analysis or parsing.

Fowle's False Orthography · · · · · · · · 45

Exercises for correction.

Page's Normal Chart · · · · · · · · · .*3 75

The elementary sounds of the language for the school-room walls.

Barber's Critical Writing Speller . . . . . 20 cts.

"The Student's Own Hand-Book of Orthography, Definitions, and Sentences, consisting of Written Exercises in the Proper Spelling, Meaning, and Use of Words." (Published 1873.) This differs from Sherwood's and other Writing Spellers in its more comprehensive character. Its blanks are adapted to writing whole sentences instead of detached words, with the proper divisions for numbering, corrections, etc. Such aids as this, like Watson's Child's Speller and Sherwood's Writing Speller, find their *raison d'être* in the postulate that the art of correct spelling is dependent upon written, and not upon spoken language, for its utility, if not for its very existence. Hence the indirectness of purely oral instruction.

ETYMOLOGY.

Smith's Complete Etymology, ... $1 25
Smith's Condensed Etymology, ... 75

Containing the Anglo-Saxon, French, Dutch, German, Welsh, Danish, Gothic, Swedish, Gaelic, Italian, Latin, and Greek Roots, and the English words derived therefrom accurately spelled, accented, and defined.

From Hon. Jno. G. McMynn, *late State Superintendent of Wisconsin.*

I wish every teacher in the country had a copy of this work.

From Prin. Wm. F. Phelps, *Minn. State Normal.*

The book is superb—just what is needed in the department of etymology and spelling.

From Prof. C. H. Verrill, *Pa. State Normal School.*

The Etymology (Smith's) which we procured of you we like much. It is the best work for the class-room we have seen.

From Hon. Edward Ballard, *Supt. of Common Schools, State of Maine.*

Many a teacher who has turned his attention to the derivation of words has rejoiced in the helps furnished by dictionaries and smaller "hand-books," where his taste could be gratified, and the labors of patient students have been available to his own improvement. A treatise on this subject, called "A Complete Etymology of the English Language," contains very much information in a small space. The author, W. W. Smith, is evidently a lover of this branch of study, and has furnished a manual of singular utility for its purpose.

DICTIONARY.

The Topical Lexicon, ... 1 75

This work is a School Dictionary, an Etymology, a compilation of synonyms, and a manual of general information. It differs from the ordinary lexicon in being arranged by topics instead of the letters of the alphabet, thus realizing the apparent paradox of a "Readable Dictionary." An unusually valuable school-book.

ENGLISH GRAMMAR.

CLARK'S DIAGRAM SYSTEM.

Clark's Easy Lessons in Language, · · · $0 35

Published 1874. Contains illustrated object-lessons of the most attractive charac. ter, and is conched in language freed as much as possible from the dry technicalities of the science.

Clark's Brief English Grammar, · · · · · 60

Published 1872. Part I. is adapted to youngest learners, and the whole forms a complete " brief course " in one volume, adequate to the wants of the common school.

Clark's Normal Grammar, · · · · · · · 1 00

Published 1870, and designed to take the place of Prof. Clark's veteran "Practical" Grammar, though the latter is still furnished upon order. The Normal is an entirely new treatise. It is a full exposition of the system as described below, with all the most recent improvements. Some of its peculiarities are—A happy blending of SYNTHESES with ANALYSES; thorough Criticisms of common errors in the use of our Language; and important improvements in the Syntax of Sentences and of Phrases.

Clark's Key to the Diagrams, · · · · · · 1 00

Clark's Analysis of the English Language, · 60

Clark's Grammatical Chart, · · · · · · · *3 75

The theory and practice of teaching grammar in American schools is meeting with a thorough revolution from the use of this system. While the old methods offer proficiency to the pupil only after much weary plodding and dull memorizing, this affords from the inception the advantage of *practical Object Teaching*, addressing the eye by means of illustrative figures; furnishes association to the memory, its most powerful aid, and diverts the pupil by taxing his ingenuity. Teachers who are using Clark's Grammar uniformly testify that they and their pupils find it the most interesting study of the school course.

Like all great and radical improvements, the system naturally met at first with much unreasonable opposition. It has not only outlived the greater part of this opposition, but finds many of its warmest admirers among those who could not at first tolerate so radical an innovation. All it wants is an impartial trial to convince the most skeptical of its merit. No one who has fairly and intelligently tested it in the school-room has ever been known to go back to the old method. A great success is already established, and it is easy to prophecy that the day is not far distant when it will be the *only system of teaching English Grammar*. As the SYSTEM is copyrighted, no other text-books can appropriate this obvious and great improvement.

Welch's Analysis of the English Sentence, · 1 25

Remarkable for its new and simple classification, its method of treating connectives, its explanations of the idioms and constructive laws of the language, etc.

10

Clark's Diagram English Grammar.

TESTIMONIALS.

From J. A. T. DUBNIN, *Principal Dubuque R. C. Academy, Iowa.*

In my opinion, it is well calculated by its system of analysis to develop those rational faculties which in the old systems were rather left to develop themselves, while the memory was overtaxed, and the pupils discouraged.

From B. A. COX, *School Commissioner, Warren County, Illinois.*

I have examined 150 teachers in the last year, and those having studied or taught Clark's System have universally stood fifty per cent. better examinations than those having studied other authors.

From M. H. B. BURKET, *Principal Masonic Institute, Georgetown, Tennessee.*

I traveled two years amusing myself in instructing (exclusively) Grammar classes with Clark's system. The first class I instructed fifty days, but found that this was more time than was required to impart a theoretical knowledge of the science. During the two years thereafter I instructed classes only *thirty* days each. Invariably I proposed that unless I prepared my classes for a more thorough, minute, and accurate knowledge of English Grammar than that obtained from the ordinary books and in the ordinary way in from one to two years, I would make no charge. I never failed in a solitary case to far exceed the hopes of my classes, and made money and character rapidly as an instructor.

From A. B. DOUGLASS, *School Commissioner, Delaware County, New York.*

I have never known a class pursue the study of it under a *live* teacher, that has not succeeded; I have never known it to have an opponent in an educated teacher who had *thoroughly* investigated it; I have never known an *ignorant* teacher to examine it; I have never known a teacher who has used it, to try any other.

From J. A. DODGE, *Teacher and Lecturer on English Grammar, Kentucky.*

We are tempted to assert that it foretells the dawn of a brighter age to our mother-tongue. Both pupil and teacher can fare sumptuously upon its contents, however highly they may have prized the manuals into which they may have been initiated, and by which their expressions have been moulded.

From W. T. CHAPMAN, *Superintendent Public Schools, Wellington, Ohio.*

I regard Clark's System of Grammar the best published. For teaching the analysis of the English Language, it surpasses any I ever used.

From F. S. LYON, *Principal South Norwalk Union School, Connecticut.*

During ten years' experience in teaching, I have used six different authors on the subject of English Grammar. I am fully convinced that Clark's Grammar is better calculated to make thorough grammarians than any other that I have seen.

From CATALOGUE OF ROHRER'S COMMERCIAL COLLEGE, *St. Louis, Missouri.*

We do not hesitate to assert, without fear of successful contradiction, that a better knowledge of the English language can be obtained by this system in six weeks than by the old methods in as many months.

From A. PICKETT, *President of the State Teachers' Association, Wisconsin.*

A thorough experiment in the use of many approved authors upon the subject of English Grammar has convinced me of the superiority of Clark. When the pupil has completed the course, he is left upon a foundation of *principle*, and not upon the *dictum* of the author.

From GEO. F. McFARLAND, *Prin. McAllisterville Academy, Juniata Co., Penn.*

At the first examination of public-school teachers by the county superintendent, when one of our student teachers commenced analyzing a sentence according to Clark, the superintendent listened in mute astonishment until he had finished, then asked what that meant, and finally, with a very knowing look, said such work wouldn't do here, and asked the applicant to parse the sentence right, and gave the lowest certificates to all who barely mentioned Clark. Afterwards, I presented him with a copy, and the next fall he permitted it to be partially used, while the third o' last fall, he openly commended the system, and appointed three of my best teachers to explain it at the two Institutes and one County Convention held since September.

☞ For further testimony of equal force, see the Publishers' Special Circular, or current numbers of the Educational Bulletin.

11

GEOGRAPHY.

NATIONAL GEOGRAPHICAL SYSTEM.

THE SERIES.

INTERMEDIATE OR ALTERNATE VOLUMES.

ACCESSORIES.

1. PRACTICAL OBJECT TEACHING. The infant scholar is first introduced to *a picture* whence he may derive notions of the shape of the earth, the phenomena of day and night, the distribution of land and water, and the great natural divisions, which mere words would fail entirely to convey to the untutored mind. Other pictures follow on the same plan, and the child's mind is called upon to grasp no idea without the aid of a pictorial illustration. Carried on to the higher books, this system culminates in Physical Geography, where such matters as climates, ocean currents, the winds, peculiarities of the earth's crust, clouds and rain, are pictorially explained and rendered apparent to the most obtuse. The illustrations used for this purpose belong to the highest grade of art.

2. CLEAR, BEAUTIFUL, AND CORRECT MAPS. In the lower numbers the maps avoid unnecessary detail, while respectively progressive, and affording the pupil new matter for acquisition each time he approaches in the constantly enlarging circle the point of coincidence with previous lessons in the more elementary books. In the Physical and Political Geography the maps embrace many new and striking features. One of the most effective of these is the new plan for displaying on each map the relative sizes of countries not represented, thus obviating much confusion which has arisen from the necessity of presenting maps in the same atlas drawn on different scales. The maps of " McNally" have long been celebrated for their superior beauty and completeness. This is the only schoolbook in which the attempt to make a *complete* atlas *also clear and distinct*, has been successful. The map *coloring* throughout the series is also noticeable. Delicate and subdued tints take the place of the startling glare of inharmonious colors which too frequently in such treatises dazzle the eyes, distract the attention, and serve to overwhelm the names of towns and the natural features of the landscape.

GEOGRAPHY—Continued.

3, THE VARIETY OF MAP EXERCISE. Starting each time from a different basis, the pupil in many instances approaches the same fact no less than *six times*, thus indelibly impressing it upon his memory. At the same time this system is not allowed to become wearisome—the extent of exercise on each subject being graduated by its relative importance or difficulty of acquisition.

4, THE CHARACTER AND ARRANGEMENT OF THE DESCRIPTIVE TEXT. The cream of the science has been carefully culled, unimportant matter rejected, elaboration avoided, and a brief and concise manner of presentation cultivated. The orderly consideration of topics has contributed greatly to simplicity. Due attention is paid to the facts in history and astronomy which are inseparably connected with, and important to the proper understanding of geography—and *such only* are admitted on any terms. In a word, the National System teaches geography as a science, pure, simple, and exhaustive.

5, ALWAYS UP TO THE TIMES. The authors of these books, editorially speaking, never sleep. No change occurs in the boundaries of countries, or of counties, no new discovery is made, or railroad built, that is not at once noted and recorded, and the next edition of each volume carries to every school-room the new order of things.

6, SUPERIOR GRADATION. This is the only series which furnishes an available volume for every possible class in graded schools. It is not contemplated that a pupil must necessarily go through every volume in succession to attain proficiency. On the contrary, *two* will suffice, but *three* are advised; and if the course will admit, the whole series should be pursued. At all events, the books are at hand for selection, and every teacher, of every grade, can find among them one *exactly suited* to his class. The best combination for those who wish to abridge the course consists of Nos. 1, 2, and 3, or where children are somewhat advanced in other studies when they commence geography, Nos. 1*, 2, and 3. Where but *two* books are admissible, Nos. 1* and 2*, or Nos. 2 and 3, are recommended.

7, FORM OF THE VOLUMES AND MECHANICAL EXECUTION. The maps and text are no longer unnaturally divorced in accordance with the time-honored practice of making text-books on this subject as inconvenient and expensive as possible. On the contrary, all map questions are to be found on the page opposite the map itself, and each book is complete in one volume. The mechanical execution is unrivalled. Paper and printing are everything that could be desired, and the binding is—A. S. Barnes and Company's.

8, MAP-DRAWING. In 1869 the system of Map-Drawing devised by Professor JEROME ALLEN was secured *exclusively* for this series. It derives its claim to originality and usefulness from the introduction of a *fixed unit of measurement* applicable to every Map. The principles being so few, simple and comprehensive, the subject of Map-Drawing is relieved of all practical difficulty. (In Nos. 2, 2*, and 3, and published separately.)

8, ANALOGOUS OUTLINES. At the same time with Map-Drawing was also introduced (in No. 2), a new and ingenious variety of Object Lessons, consisting of a comparison of the outlines of countries with familiar objects pictorially represented.

13

MONTEITH'S INDEPENDENT COURSE.

Elementary Geography (published 1874) . . $0 80

Comprehensive Geography (with 103 Maps) . 1 60

☞ These volumes are not revisions of old works—not an addition to any series—but are entirely new productions—each by itself complete, independent, comprehensive, yet simple, brief, cheap, and popular; or, taken together, the most admirable "series" ever offered for a common-school course. They present the following features, skillfully interwoven—the student learning all about one country at a time.

LOCAL GEOGRAPHY, or the Use of Maps. Important features of the Maps are the coloring of States as objects, and the ingenious system for laying down a much larger number of names for reference than are found on any other Maps of same size—and without crowding.

PHYSICAL GEOGRAPHY, or the Natural Features of the Earth, illustrated by the original and striking *Relief Maps,* being bird's-eye views or photographic pictures of the Earth's surface.

DESCRIPTIVE GEOGRAPHY, including the Physical; with some account of Governments, and Races, Animals, etc.

HISTORICAL GEOGRAPHY, or a brief summary of the salient points of history, explaining the present distribution of nations, origin of geographical names, etc.

MATHEMATICAL GEOGRAPHY, including Astronomical, which describes the Earth's position and character among planets; also the Zones, Parallels, etc.

COMPARATIVE GEOGRAPHY, or a system of analogy, connecting new lessons with the previous ones. Comparative sizes and latitudes are shown on the margin of each Map, and all countries are measured in the "*frame of Kansas.*"

TOPICAL GEOGRAPHY, consisting of questions for review, and testing the student's general and specific knowledge of the subject, with suggestions for *Geographical Compositions.*

ANCIENT GEOGRAPHY. A section devoted to this subject, with Maps, will be appreciated by teachers. It is seldom taught in our common schools, because it has heretofore required the purchase of a separate book.

GRAPHIC GEOGRAPHY, or Map-Drawing by Allen's "Unit of Measurement" system (now almost universally recognized as without a rival) is introduced throughout the lessons, and not as an appendix.

CONSTRUCTIVE GEOGRAPHY, or Globe-Making. With each book a set of Map Segments is furnished, with which each student may make his own Globe by following the directions given.

RAILROAD GEOGRAPHY, with a grand Map illustrating routes of travel in the United States. Also, a "Tour in Europe."

14

The National System of Geography,

BY MONTEITH & McNALLY.

ITS RECORD.

These popular text-books have been adopted, by official authority, for the schools of the following States and Cities—in most cases for *exclusive* and uniform use.

STATES.

California,	Vermont,	Florida,
Missouri,	Iowa,	Minnesota,
Alabama,	Louisiana,	North Carolina,
Tennessee,	Oregon,	Kansas,
Texas,	Arkansas,	Mississippi.

CITIES.

New York City,	Louisville,	Nashville,	Portland,
Brooklyn,	Newark,	Utica,	Savannah,
New Orleans,	Milwaukee,	Wilmington,	Indianapolis,
Buffalo,	Charleston,	Trenton,	Springfield,
Richmond,	Rochester,	Norfolk,	Wheeling,
Jersey City,	Mobile,	Norwich,	Toledo,
Hartford,	Syracuse,	Lockport,	Bridgeport,
Worcester,	Memphis,	Dubuque,	St. Paul,
San Francisco,	Salt Lake City,	Galveston,	Vicksburg,
&c.	&c.	&c.	&c.

STANDARD WALL MAPS.

By JAMES MONTEITH.

Monteith's School Maps, 8 Numbers, per set **$20 00*

The "School Series" includes the Hemispheres (2 Maps), United States, North America, South America, Europe, Asia, Africa.—Price. $2.50 each.

Each Map is 28 × 34 inches, beautifully colored, has the names all laid down, and is substantially mounted on canvas with rollers.

Monteith's Grand Maps, 7 Numbers, per set (in locked box) **35 00*

The "Grand Series" includes the Hemispheres (1 Map), United States, South America, Europe, Asia, Africa, The World on Mercator's Projection.—Price, $5.00 each. Size 42 × 52 inches, names laid down, colored, mounted, &c., like the School Series.

Monteith & McNally's National Geographies.

CRITICAL OPINIONS.

From R. A. ADAMS, *Member of Board of Education, New York.*

I have found, by examination of the Book of Supply of our Board, that considerably the largest number of any series now used in our public schools is the National, by Monteith and McNally.

From BRO. PATRICK, *Chief Provincial of the Vast Educational Society of the* CHRISTIAN BROTHERS *in the United States.*

Having been convinced for some time past that the series of Geographies in use in our schools were not giving satisfaction, and came far short of meeting our most reasonable expectations, I have felt it my imperative duty to examine into this matter, and see if a remedy could not be found.

Copies of the different Geographies published in this country have been placed at our command for examination. On account of other pressing duties we have not been able to give as much time to the investigation of all these different series as we could have desired ; yet we have found enough to convince us that there are many others better than those we are now using ; but we cheerfully give our most decided preference, above all others, to the National Series, by Monteith & McNally.

Their easy gradation, their thoroughly practical and independent character, their comprehensive completeness as a full and accurate system, the wise discrimination shown in the selection of the subject matter, the beautiful and copious illustrations, the neat cut type, the general execution of the works, and *other excellencies*, will commend them to the friends of education everywhere.

From the "HOME MONTHLY," *Nashville, Tenn.*

MONTEITH'S AND McNALLY'S GEOGRAPHIES.—Geography is so closely connected with Astronomy, History, Ethnology, and Geology, that it is difficult to define its limits in the construction of a text-book. If the author confines himself strictly to a description of the earth's surface, his book will be dry, meager, and unintelligible to a child. If, on the other hand, he attempts to give information on the cognate sciences, he enters a boundless field, and may wander too far. It seems to us that the authors of the series before us have hit on the happy medium between too much and too little. *The First Lessons*, by applying the system of object-teaching, renders the subject so attractive that a child, just able to read, may become deeply interested in it. The second book of the course enlarges the view, but still keeps to the maps and simple descriptions. Then, in the third book. we have Geography combined with History and Astronomy. A general view of the solar system is presented, so that the pupil may understand the earth's position on the map of the heavens. The first part of the fourth book treats of Physical Geography, and contains a vast amount of knowledge compressed into a small space. It is made bright and attractive by beautiful pictures and suggestive illustrations, on the principle of object-teaching. The maps in the second part of this volume are remarkably clear, and the map exercises are copious and judicious. In the fifth and last volume of the series, the whole subject is reviewed and systematized. This is strictly a Geography. Its maps are beautifully engraved and clearly printed. The map exercises are full and comprehensive. In all these books the maps, questions and descriptions are given in the same volume. In most geographies there are too many details and minute descriptions—more than any child out of purgatory ought to be required to learn. The power of memory is overstrained ; there is confusion—no clearly defined idea is formed in the child's mind. But in these books, in brief, pointed descriptions, and constant use of bright, accurate maps, the whole subject is photographed on the mind.

16

MATHEMATICS.

DAVIES' NATIONAL COURSE.

ARITHMETIC,

SLATED.

1. Davies' Primary Arithmetic,. . . . .$ 25 $ 32
2. Davies' Intellectual Arithmetic, ... 40 48
3. Davies' Elements of Written Arithmetic,. . . , . 50 60
4. Davies' Practical Arithmetic, ... 90 1 00
   Key to Practical Arithmetic, ... 90
5. Davies' University Arithmetic,. 1 40 1 50
   Key to University Arithmetic, ... *1 40

ALGEBRA.

1. Davies' New Elementary Algebra, ... *1 25 1 35
   Key to Elementary Algebra, ... *1 25
2. Davies' University Algebra,. 1 50 1 60
   Key to University Algebra,. *1 50
3. Davies' New Bourdon's Algebra, ... 2 25 2 38
   Key to Bourdon's Algebra, ... *2 25

GEOMETRY.

1. Davies' Elementary Geometry and Trigonometry, 1 40 1 50
2. Davies' Legendre's Geometry, ... 2 25 2 38
3. Davies' Analytical Geometry and Calculus,. . 2 50 2 63
4. Davies' Descriptive Geometry, ... 2 75 2 88
5. Davies' New Calculus, ... 2 00

MENSURATION.

1. Davies' Practical Mathematics and Mensuration,. 1 50 1 60
2. Davies' Elements of Surveying, ... 2 50 2 63
3. Davies' Shades, Shadows, and Perspective,. . . 3 75 3 88

MATHEMATICAL SCIENCE.

Davies' Grammar of Arithmetic,. . . . . * 50
Davies' Outlines of Mathematical Science, ... *1 00
Davies' Nature and Utility of Mathematics, 8vo, *2 00, 12mo, *1 50
Davies' Metric System, ... *1 50
Davies & Peck's Dictionary of Mathematics, ... *5 00
Davies' Foundations Mathematical Science, ... * 25

17

MATHEMATICS—Continued.

ARITHMETICAL EXAMPLES.

Reuck's Examples in Denominate Numbers $ 50
Reuck's Examples in Arithmetic 1 00

These volumes differ from the ordinary arithmetic in their peculiarly *practical* character. They are composed mainly of examples, and afford the most severe and thorough discipline for the mind. While a book which should contain a complete treatise of theory and practice would be too cumbersome for every-day use, the insufficiency of *practical* examples has been a source of complaint.

HIGHER MATHEMATICS.

Church's Elements of Calculus 2 50
Church's Analytical Geometry 2 50
Church's Descriptive Geometry, with Shades,
    Shadows, and Perspective 4 00

These volumes constitute the "West Point Course" in their several departments.

Courtenay's Elements of Calculus 3 00

A werk especially popular at the South.

Hackley's Trigonometry 2 50

With applications to navigation and surveying, nautical and practical geometry and geodesy.

Peck's Analytical Geometry 1 75
Peck's Practical Calculus 1 75

APPLIED MATHEMATICS.

Peck's Ganot's Popular Physics 1 75
Peck's Elements of Mechanics 2 00
Peck's Practical Calculus 1 75
Peck's Analytical Geometry, ... 1 75

Prof. W. G. Peck, of Columbia College, has designed the first of these works for the ordinary wants of schools in the department of Natural Philosophy. The other volumes are the briefest treatises on those subjects now published. Their methods are purely practical, and unembarrassed by the details which rather confuse than simplify science.

SLATED ARITHMETICS.

This consists of the application of an artificially slated surface to the inner cover of a book, with flap of the same opening outward, so that students may refer to the book and use the slate at one and the same time, and as though the slate were detached. When folded up, the slate preserves examples and memoranda til' needed. The material used is as durable as the stone slate. The additional cost of books thus improved is trifling.

Davies' National Course of Mathematics.

TESTIMONIALS.

From L. Van Bokkelen, State Superintendent Public Instruction, Maryland.

The series of Arithmetics edited by Prof. Davies, and published by your firm, have been used for many years in the schools of several counties, and the city of Baltimore, and have been approved by teachers and commissioners.

Under the law of 1865, establishing a uniform system of Free Public Schools, these Arithmetics were unanimously adopted by the State Board of Education, after a careful examination, and are now used in all the Public Schools of Maryland.

These facts evidence the high opinion entertained by the School Authorities of the value of the series theoretically and practically.

From Horace Webster, President of the College of New York.

The undersigned has examined, with care and thought, several volumes of Davies' Mathematics, and is of the opinion that, as a whole, it is the most complete and best course for Academic and Collegiate instruction, with which he is acquainted.

From David N. Camp, State Superintendent of Common Schools, Connecticut.

I have examined Davies' Series of Arithmetics with some care. The language is clear and precise; each principle is thoroughly analyzed, and the whole so arranged as to facilitate the work of instruction. Having observed the satisfaction and success with which the different books have been used by eminent teachers, it gives me pleasure to commend them to others.

From J. O. Wilson, Chairman Committee on Text-Books, Washington, D. C.

I consider Davies' Arithmetics decidedly superior to any other series, and in this opinion I am sustained, I believe, by the entire Board of Education and Corps of Teachers in this city, where they have been used for several years past.

From John L. Campbell, Professor of Mathematics, Wabash College, Indiana.

A proper combination of abstract reasoning and practical illustration is the chief excellence in Prof. Davies' Mathematical works. I prefer his Arithmetics, Algebras, Geometry and Trigonometry to all others now in use, and cordially recommend them to all who desire the advancement of sound learning.

From Major J. H. Whittlesey, Government Inspector of Military Schools.

Be assured, I regard the works of Prof. Davies, with which I am acquainted, as by far the best text-books in print on the subjects which they treat. I shall certainly encourage their adoption wherever a word from me may be of any avail.

From T. McC. Ballantine, Prof. Mathematics Cumberland College, Kentucky.

I have long taught Prof. Davies' Course of Mathematics, and I continue to like their working.

From John McLean Bell, B. A., Prin. of Lower Canada College.

I have used Davies' Arithmetical and Mathematical Series as text-books in the schools under my charge for the last six years. These I have found of great efficacy in exciting, invigorating, and concentrating the intellectual faculties of the young.

Each treatise serves as an introduction to the next higher, by the similarity of its reasonings and methods; and the student is carried forward, by easy and gradual steps, over the whole field of mathematical inquiry, and that, too, in a *shorter* time than is usually occupied in mastering a single department. I sincerely and heartily recommend them to the attention of my fellow-teachers in Canada.

From D. W. Steele, Prin. Philekoian Academy, Cold Springs, Texas.

I have used Davies' Arithmetics till I know them nearly by heart. A better series of school-books never were published. I have recommended them until they are now used in all this region of country.

A large mass of similar "Opinions" may be obtained by addressing the publishers for special circular for Davies' Mathematics. New recommendations are published in current numbers of the *Educational Bulletin.*

DAVIES' NATIONAL COURSE of MATHEMATICS.

ITS RECORD.

In claiming for this series the first place among American text-books, of what ever class, the Publishers appeal to the magnificent record which its volumes have earned during the *thirty-five years* of Dr. Charles Davies' mathematical labors. The unremitting exertions of a life-time have placed *the modern series* on the same proud eminence among competitors that each of its predecessors has successively enjoyed in a course of constantly improved editions, now rounded to their perfect fruition—for it seems almost that this science is susceptible of no further demonstration.

During the period alluded to, many authors and editors in this department have started into public notice, and by borrowing ideas and processes original with Dr. Davies, have enjoyed a brief popularity, but are now almost unknown. Many of the series of to-day, built upon a similar basis, and described as "modern books," are destined to a similar fate; while the most far-seeing eye will find it difficult to fix the time, on the basis of any data afforded by their past history, when these books will cease to increase and prosper, and fix a still firmer hold on the affection of every educated American.

One cause of this unparalleled popularity is found in the fact that the enterprise of the author did not cease with the original completion of his books. Always a practical teacher, he has incorporated in his text-books from time to time the advantages of every improvement in methods of teaching, and every advance in science. During all the years in which he has been laboring, he constantly submitted his own theories and those of others to the practical test of the class-room —approving, rejecting, or modifying them as the experience thus obtained might suggest. In this way he has been able to produce an almost perfect series of class-books, in which every department of mathematics has received minute and exhaustive attention.

Nor has he yet retired from the field. Still in the prime of life, and enjoying a ripe experience which no other living mathematician or teacher can emulate, his pen is ever ready to carry on the good work, as the progress of science may demand. Witness his recent exposition of the "Metric System," which received the official endorsement of Congress, by its Committee on Uniform Weights and Measures.

DAVIES' SYSTEM IS THE ACKNOWLEDGED NATIONAL STANDARD FOR THE UNITED STATES, for the following reasons:—

1st. It is the basis of instruction in the great national schools at West Point and Annapolis.

2d. It has received the *quasi* endorsement of the National Congress.

3d. It is exclusively used in the public schools of the National Capital.

4th. The officials of the Government use it as authority in all cases involving mathematical questions.

5th. Our great soldiers and sailors commanding the national armies and navies were educated in this system. So have been a majority of eminent scientists in this country. All these refer to "Davies" as authority.

6th. A larger number of American citizens have received their education from this than from any other series.

7th. The series has a larger circulation throughout the whole country than any other, being *extensively used in every State in the Union.*

MATHEMATICS—Continued.

PECK'S ARITHMETICS.

By the Prof. of Mathematics at Columbia College, New York.

1. Peck's First Lessons in Numbers, · · · $0 25

Embracing all that is usually included in what are called Primary and Intellectual Arithmetics; proceeding gradually from object lessons to abstract numbers; developing Addition and Subtraction simultaneously: with other attractive novelties.

2. Peck's Manual of Practical Arithmetic, · 50

An excellent "Brief" course, conveying a sufficient knowledge of Arithmetic for ordinary business purposes.

It is thoroughly "practical," because the author believes the Theory cannot be studied with advantage until the pupil has acquired a certain facility in combining numbers, which can only be had by practice.

3. Peck's Complete Arithmetic, · · · · · 90

The whole subject—theory and practice—presented within very moderate limits. This author's most remarkable faculty of mathematical treatment is comprehended in three words: System, Conciseness, Lucidity. The directness and simplicity of this work cannot be better expressed than in the words of a correspondent who adopted the book at once, because, as he said, it is "free from that *juggling with numbers*" practiced by many authors.

From the "Galaxy," New York.

In the "Complete Arithmetic" each part of the subject is logically developed. First are given the necessary definitions; second, the explanations of such signs (if any) as are used; third, the principles on which the operation depends; fourth, an exemplification of the manner in which the operation is performed, which is so conducted that the reason of the rule which is immediately thereafter deduced is made perfectly plain; after which follow numerous graded examples and corresponding practical problems. All the parts taken together are arranged in logical order. The subject is treated as a whole, and not as if made up of segregated parts. It may seem a simple remark to make that (for example) addition is in principle one and the same everywhere, whether employed upon simple or compound numbers, fractions, etc., the only difference being in the *unit* involved; but the number of persons who understand this practically, compared to the number who have studied arithmetic, is not very great. The student of the "Complete Arithmetic" cannot fail to understand it. All the principles of the science are presented within moderate limits. Superfluity of matter—to supplement defective definitions. to make clear faulty demonstrations and rules expressed either inaccurately or obscurely, to make provision for a multiplicity of cases for which no provision is requisite—has been carefully avoided. The definitions are plain and concise; the principles are stated clearly and accurately; the demonstrations are full and complete; the rules are perspicuous and comprehensive; the illustrative examples are abundant and well fitted to familiarize the student with the application of principles to the problems of science and of every-day life.

☞ The Definitions constitute the power of the book. We have never seen them excelled for clearness and exactness.—*Iowa School Journal.*

PENMANSHIP.

Beers' System of Progressive Penmanship.
Per dozen $1 68

This "round hand" system of Penmanship in twelve numbers, commends itself by its simplicity and thoroughness. The first four numbers are primary books. Nos. 5 to 7, advanced books for boys. Nos. 8 to 10, advanced books for girls. Nos. 11 and 12, ornamental penmanship. These books are printed from steel plates (engraved by McLees), and are unexcelled in mechanical execution. Large quantities are annually sold.

Beers' Slated Copy Slips, per set *50

All beginners should practice, for a few weeks, slate exercises, familiarizing them with the form of the letters, the motions of the hand and arm, &c., &c. These copy slips, 32 in number, supply all the copies found in a complete series of writing-books, at a trifling cost.

Payson, Dunton&Scribner's Copy-B'ks.P.doz.,*1 80

The National System of Penmanship, in three distinct series—(1) Common School Series, comprising the first six numbers; (2) Business Series, Nos. 8, 11, and 12; (3) Ladies' Series, Nos. 7, 9, and 10.

Fulton & Eastman's Chirographic Charts,*3 75

To embellish the school room walls, and furnish class exercise in the elements of Penmanship.

Payson's Copy-Book Cover, per hundred .*4 00

Protects every page except the one in use, and furnishes "lines" with proper slope for the penman, under. Patented.

National Steel Pens, Card with all kinds . . . *15

Pronounced by competent judges the perfection of American-made pens, and superior to any foreign article.

SCHOOL SERIES.		Index Pen, per gross . . . 75
School Pen, per gross, . .$ 60		BUSINESS SERIES.
Academic Pen, do . . 63		Albata Pen, per gross, . . 40
Fine Pointed Pen, per gross 70		Bank Pen, do . . 10
POPULAR SERIES.		Empire Pen, do . . 70
Capitol Pen, per gross, . . 1 00		Commercial Pen, per gross . 60
do do pr. box of 2 doz. 25		Express Pen, do . 75
Bullion Pen (imit. gold) pr. gr. 75		Falcon Pen, do . 70
Ladies' Pen do 63		Elastic Pen, do . 75

Stimpson's Scientific Steel Pen, per gross .*2 00

One forward and two backward arches, ensuring great strength, well-balanced elasticity, evenness of point, and smoothness of execution. One gross in twelve contains a Scientific Gold Pen.

Stimpson's Ink-Retaining Holder, per doz. .*2 00

A simple apparatus, which does not get out of order, withholds at a single dip as much ink as the pen would otherwise realize from a dozen trips to the inkstand, which it supplies with moderate and easy flow.

Stimpson's Gold Pen, $3 00; with Ink Retainer*4 50
Stimpson's Penman's Card, ...* 50

One dozen Steel Pens (assorted points) and Patent Ink-retaining Pen holder.

HISTORY.

Monteith's Youth's History, ... $ 75

A History of the United States for beginners. It is arranged upon the catechetical plan, with illustrative maps and engravings, review questions, dates iu parentheses (that their study may be optional with the younger class of learners), and interesting biographical Sketches of all persons who have been prominently identified with the history of our country.

Willard's United States, School edition, . . . 1 40
Do.   do.   University edition, . 2 25

The plan of this standard work is chronologically exhibited in front of the title-page; the Maps and Sketches are found useful assistants to the memory, and dates, usually so difficult to remember, are so systematically arranged as in a great degree to obviate the difficulty. Candor, impartiality, and accuracy, are the distinguishing features of the narrative portion.

Willard's Universal History, ... 2 25

The most valuable features of the "United States" are reproduced in this. The peculiarities of the work are its great conciseness and the prominence given to the chronological order of events. The margin marks each successive era with great distinctness, so that the pupil retains not only the event but its time, and thus fixes the order of history firmly and usefully in his mind. Mrs. Willard's books are constantly revised, and at all times written up to embrace important historical events of recent date.

Berard's History of England, ... 1

By an authoress well known for the success of her History of the United States. The social life of the English people is felicitously interwoven, as in fact, with the civil and military transactions of the realm.

Ricord's History of Rome, ... 1 75

Possesses the charm of an attractive romance. The Fables with which this history abounds are introduced in such a way as not to deceive the inexperienced, while adding materially to the value of the work as a reliable index to the character and institutions, as well as the history of the Roman people.

Hanna's Bible History, ... 1 25

The only compendium of Bible narrative which affords a connected and chronological view of the important events there recorded, divested of all superfluous detail.

Summary of History, Complete 60
American History, $0 40.   French and Eng. Hist.   35

A well proportioned outline of leading events, condensing the substance of the more extensive text-book in common use into a series of statements so brief, that every word may be committed to memory, and yet so comprehensive that it presents an accurate though general view of the whole continuous life of nations.

Marsh's Ecclesiastical History, ... 2 00
Questions to ditto, ... 75

Affording the History of the Church in all ages, with accounts of the pagan world during Biblical periods, and the character, rise, and progress of all Religions, as well as the various sects of the worshipers of Christ. The work is entirely non-sectarian, though strictly catholic.

Mill's History of the Jews, ... 1 75

BARNES' ONE-TERM HISTORY.

A Brief History of the United States, · · .$1 50

This is probably the MOST ORIGINAL SCHOOL-BOOK published for many years, in any department. A few of its claims are the following:

1. Brevity.—The text is complete for Grammar School or intermediate classes, in 290 12mo pages, large type. It may readily be completed, if desired, in one term of study.

2. Comprehensiveness.—Though so brief, this book contains the pith of all the wearying contents of the larger manuals, and a great deal more than the memory usually retains from the latter.

3. Interest has been a prime consideration. Small books have heretofore been bare, full of dry statistics, unattractive. This one is charmingly written, replete with anecdote, and brilliant with illustration.

4. Proportion of Events.—It is remarkable for the discrimination with which the different portions of our history are presented according to their importance. Thus the older works being already large books when the civil war took place, give it less space than that accorded to the Revolution.

5. Arrangement.—In six epochs, entitled respectively, Discovery and Settlement, the Colonies, the Revolution, Growth of States, the Civil War, and Current Events.

6. Catch Words.—Each paragraph is preceded by its leading thought in prominent type, standing in the student's mind for the whole paragraph.

7. Key Notes.—Analogous with this is the idea of grouping battles, etc., about some central event, which relieves the sameness so common in such descriptions, and renders each distinct by some striking peculiarity of its own.

8. Foot Notes.—These are crowded with interesting matter that is not strictly a part of history proper. They may be learned or not, at pleasure. They are certain in any event to be read.

9. Biographies of all the leading characters are given in full in foot-notes.

10. Maps.—Elegant and distinct Maps from engravings on copper-plate, and beautifully colored, precede each epoch, and contain all the places named.

11. Questions are at the back of the book, to compel a more independent use of the text. Both text and questions are so worded that the pupil must give intelligent answers IN HIS OWN WORDS. "Yes" and "No" will not do.

12. Historical Recreations.—These are additional questions to test the student's knowledge, in review, as : "What trees are celebrated in our history?" "When did a fog save our army?" "What Presidents died in office?" "When was the Mississippi our western boundary?" "Who said, 'I would rather be right than President?'" etc.

13. The Illustrations, about seventy in number, are the work of our best artists and engravers, produced at great expense. They are vivid and interesting, and mostly upon subjects never before illustrated in a school-book.

14. Dates.—Only the leading dates are given in the text, and these are so associated as to assist the memory, but at the head of each page is the date of the event first mentioned, and at the close of each epoch a summary of events and dates.

15. The Philosophy of History is studiously exhibited—the causes and effects of events being distinctly traced and their interconnection shown.

16. Impartiality.—All sectional, partisan, or denominational views are avoided. Facts are stated after a careful comparison of all authorities without the least prejudice or favor.

17. Index.—A verbal index at the close of the book perfects it as a work of reference.

It will be observed that the above are all particulars in which School Histories have been signally defective, or altogether wanting. Many other claims to favor it shares in common with its predecessors.

HISTORY—Continued.

Hunter's Historical Games, with cards . . . $0 75

An invaluable accompaniment for the text-book, by way of stimulating interest in the Class ; affording, at once, Amusement and Instruction.

SOME TESTIMONIALS FOR BARNES' BRIEF HISTORY.

From HON. J. M. McKENZIE, *Supt. Pub. Inst., Nebraska.*
I have examined your "Brief History of the United States," and like it *real well ;* and were I teaching a graded school, I think I should use it as a text-book.

From HON. H. B. WILSON, *Supt. Pub. Inst., Minnesota.*
I have read with much interest the "One-Term History of the United States." I am much pleased with it. In my judgment, it contains all of the United States history that the majority of pupils in our common schools can spare time to study.

From PRES. EDWARD BROOKS, *Millersville State Normal School, Pa.*
It is a work that will be a favorite with teachers and pupils. Its scope and style especially adapt it for use in our public schools. I cordially commend it to teachers desiring to introduce an interesting and practical text-book upon this subject.

From PRES. BARKER, *Buffalo State Normal School, N. Y.*
In the copy of your "Brief History," before me, the important items to be learned in history seem most ingeniously brought out and kept in the foreground. These items are *time. persons, places,* and *events.* It has the appearance of an exceedingly fresh and systematic work. I think I shall put it into my classes.

From PROF. WM. F. ALLEN, *State Univ. of Wisconsin.*
I think the author of the new "Brief History of the United States" has been very successful in combining brevity with sufficient fullness and interest. *Particularly,* he has avoided the excessive number of names and dates that most histories contain. Two features that I like *very much* are the *anecdotes* at the foot of the page and the "*Historical Recreations*" in the Appendix. The latter, I think, is quite a *new* feature, and the other is *very* well executed.

From S. G. WRIGHT, *Assist.-Supt. Pub. Inst., Kansas.*
It is with extreme pleasure we submit our recommendation of the "Brief History of the United States." It meets the needs of young and older children, combining concision with perspicuity, and if "brevity is the soul of wit," this "Brief History" contains not only that well-chosen ingredient, but wisdom sufficient to enlighten those students who are wearily longing for a "new departure" from certain old and uninteresting presentations of fossilized writers. We congratulate a progressive public upon a progressive book.

From HON. NEWTON BATEMAN. *Supt. Pub. Inst., Illinois.*
Barnes' One-Term History of the United States is an exceedingly attractive and spirited little book. Its claim to several new and valuable features seems well founded. Under the form of six well-defined Epochs, the History of the United States is traced tersely, yet pithily, from the earliest times to the present day. A good map precedes each epoch, whereby the history and geography of the period may be studied together, *as they always should be.* The syllabus of each paragraph is made to stand in such bold relief, by the use of large, heavy type, as to be of much *mnemonic* value to the student. The book is written in a sprightly and piquant style. the interest never flagging from beginning to end—a rare and difficult achievement in works of this kind.

From the "Chicago Schoolmaster" (Editorial).
A thorough examination of Barnes' Brief History of the United States brings the examiner to the conclusion that it is a superior book in almost every respect. The book is neat in form, and of good material. The type is clear, large, and distinct. The facts and dates are correct. The arrangement of topics is just the thing needed in a history text-book. By this arrangement the pupil can see at once what he is expected to do. The topics are well selected, embracing the leading ideas or principal events of American history. . . . The book as a whole is much superior to any I have examined. So much do I think this, that I have ordered it for my class, and shall use it in my school.   (Signed)   B. W. BAKER.

Baker's Brief History of Texas, · · · · · $1 25

DRAWING.

Chapman's American Drawing Book, . . .*$6 00

The standard American text-book and authority in all branches of art. A compilation of art principles. A manual for the amateur, and basis of study for the professional artist. Adapted for schools and private instruction.

CONTENTS.—"Any one who can Learn to Write can Learn to Draw."—Primary Instruction in Drawing.—Rudiments of Drawing the Human Head.—Rudiments in Drawing the Human Figure.—Rudiments of Drawing.—The Elements of Geometry.—Perspective.—Of Studying and Sketching from Nature.—Of Painting.—Etching and Engraving.—Of Modeling.—Of Composition—Advice to the American Art-Student. The work is of course magnificently illustrated with all the original designs.

Chapman's Elementary Drawing Book, . . 1 50

A Progressive Course of Practical Exercises, or a text-book for the training of the eye and hand. It contains the elements from the larger work, and a copy should be in the hands of every pupil; while a copy of the "American Drawing Book," named above, should be at hand for reference by the class.

The Little Artist's Portfolio, ... *50

25 Drawing Cards (progressive patterns), 25 Blanks, and a fine Artist's Pencil, all in one neat envelope.

Clark's Elements of Drawing, ...*1 00

A complete course in this graceful art, from the first rudiments of outline to the finished sketches of landscape and scenery.

Fowle's Linear and Perspective Drawing, . *60

For the cultivation of the eye and hand, with copious illustrations and directions for the guidance of the unskilled teacher.

Monk's Drawing Books—Six Numbers, per set, *2 25

Each book contains *eleven* large patterns, with opposing blanks. No. 1. Elementary Studies. No. 2. Studies of Foliage. No. 3. Landscapes. No. 4. Animals, I. No. 5. Animals, II. No. 6. Marine Views, etc.

Allen's Map-Drawing, . . . 25 cts.; Scale, 25

This method introduces a new era in Map-Drawing, for the following reasons:—1. It is a system. This is its greatest merit.—2. It is easily understood and taught.—3. The eye is trained to exact measurement by the use of a scale.—4. By no special effort of the memory, distance and comparative size are fixed in the mind.—5. It discards useless construction of lines.—6. It can be taught by any teacher, even though there may have been no previous practice in Map-Drawing.—7. Any pupil old enough to study Geography can learn by this System, in a short time, to draw accurate maps.—8. The System is not the result of theory, but comes directly from the school-room. It has been thoroughly and successfully tested there, with all grades of pupils.—9. It is economical, as it requires no mapping plates. It gives the pupil the ability of rapidly drawing accurate maps.

Ripley's Map-Drawing, ... 1 25

Based on the Circle. One of the most efficient aids to the acquirement of a knowledge of Geography is the practice of map-drawing. It is useful for the same reason that the best exercise in orthography is the *writing* of difficult words. Sight comes to the aid of hearing, and a double impression is produced upon the memory. Knowledge becomes less mechanical and more intuitive. The student who has sketched the outlines of a country, and dotted the important places, is little likely to forget either. The impression produced may be compared to that of a traveller who has been over the ground, while more comprehensive and accurate in detail.

BOOK-KEEPING.

Folsom's Logical Book-keeping, ... $2 00
Folsom's Blanks to Book-keeping, ... *4 50

This treatise embraces the interesting and important discoveries of Prof. Folsom (of the Albany "Bryant & Stratton College"), the partial enunciation of which in lectures and otherwise has attracted so much attention in circles interested in commercial education.

After studying business phenomena for many years, he has arrived at the positive laws and principles that underlie the whole subject of Accounts; finds that the science is based in *Value* as a generic term; that value divides into *two classes* with varied species; that all the exchanges of values are reducible to nine equations; and that all the results of all these exchanges are limited to *thirteen* in number.

As accounts have been universally taught hitherto, without setting out from a radical analysis or definition of values, the science has been kept in great obscurity, and been made as difficult to impart as to acquire. On the new theory, however, these obstacles are chiefly removed. In reading over the first part of it, in which the governing laws and principles are discussed, a person with ordinary intelligence will obtain a fair conception of the *double entry* process of accounts. But when he comes to study thoroughly these laws and principles as there enunciated, and works out the examples and memoranda which elucidate the *thirteen results* of business, the student will neither fail in readily acquiring the science as it is, nor in becoming able intelligently to apply it in the interpretation of business.

Smith & Martin's Book-keeping, ... 1 25
Smith & Martin's Blanks, ... *60

This work is by a practical teacher and a practical book-keeper. It is of a thoroughly popular class, and will be welcomed by every one who loves to see theory and practice combined in an easy, concise, and methodical form.

The Single Entry portion is well adapted to supply a want felt in nearly all other treatises, which seem to be prepared mainly for the use of wholesale merchants, leaving retailers, mechanics, farmers, etc., who transact the greater portion of the business of the country, without a guide. The work is also commended, on this account, for general use in Young Ladies' Seminaries, where a thorough grounding in the simpler form of accounts will be invaluable to the future housekeepers of the nation.

The treatise on Double Entry Book-keeping combines all the advantages of the most recent methods, with the utmost simplicity of application, thus affording the pupil all the advantages of actual experience in the counting-house, and giving a clear comprehension of the entire subject through a judicious course of mercantile transactions.

The shape of the book is such that the transactions can be presented as in actual practice; and the simplified form of Blanks— three in number—adds greatly to the ease experienced in acquiring the science.

NATURAL SCIENCE.

FAMILIAR SCIENCE.

Norton & Porter's First Book of Science, · $1 75

By eminent Professors of Yale College. Contains the principles of Natural Philosophy, Astronomy, Chemistry, Physiology, and Geology. Arranged on the Catechetical plan for primary classes and beginners.

Chambers' Treasury of Knowledge, · · · · 1 25

Progressive lessons upon—*first*, common things which lie most immediately around us, and first attract the attention of the young mind; *second*, common objects from the Mineral, Animal, and Vegetable kingdoms, manufactured articles, and miscellaneous substances; *third*, a systematic view of Nature under the various sciences. May be used as a Reader or Text-book.

NATURAL PHILOSOPHY.

Norton's First Book in Natural Philosophy, 1 00

By Prof. NORTON, of Yale College. Designed for beginners. Profusely illustrated and arranged on the Catechetical plan.

Peck's Ganot's Course of Nat. Philosophy, · 1 75

The standard text-book of France, Americanized and popularized by Prof. PECK, of Columbia College. The most magnificent system of illustration ever adopted in an American school-book is here found. For intermediate classes.

Peck's Elements of Mechanics, · · · · · · 2 00

A suitable introduction to Bartlett's higher treatises on Mechanical Philosophy, and adequate in itself for a complete academical course.

Bartlett's SYNTHETIC, AND ANALYTIC, Mechanics, · each 5 00

Bartlett's Acoustics and Optics, · · · · · 3 50

A system of Collegiate Philosophy, by Prof. BARTLETT, of West Point Military Academy.

Steele's 14 Weeks Course in Philos. (see p. 34) 1 50

Steele's Philosophical Apparatus, · · · · *125 00

Adequate to performing the experiments in the ordinary text-books. The articles will be sold separately, if desired. See special circular for details.

GEOLOGY.

Page's Elements of Geology, · · · · · · · 1 25

A volume of Chambers' Educational Course. Practical, simple, and eminently calculated to make the study interesting.

Emmons' Manual of Geology, · · · · · · 1 25

The first Geologist of the country has here produced a work worthy of his reputation.

Steele's 14 Weeks Course (see p. 34) · · · · · 1 50

Steele's Geological Cabinet, · · · · · · · *40 00

Containing 125 carefully selected specimens. In four parts. Sold separately, if desired. See circular for details.

Peck's Ganot's Popular Physics.

TESTIMONIALS.

From PROF. ALONZO COLLIN, *Cornell College, Iowa.*

I am pleased with it. I have decided to introduce it as a text-book.

From H. F. JOHNSON, *President Madison College, Sharon, Miss.*

I am pleased with Peck's Ganot, and think it a magnificent book.

From PROF. EDWARD BROOKS, *Pennsylvania State Normal School.*

So eminent are its merits, that it will be introduced as the text-book upon elementary physics in this institution.

From H. H. LOCKWOOD, *Professor Natural Philosophy U. S. Naval Academy.*

I am so pleased with it that I will probably add it to a course of lectures given to the midshipmen of this school on physics.

From GEO. S. MACKIE, *Professor Natural History University of Nashville, Tenn.*

I have decided on the introduction of Peck's Ganot's Philosophy, as I am satisfied that it is the best book for the purposes of my pupils that I have seen, combining simplicity of explanation with elegance of illustration.

From W. S. MCRAE, *Superintendent Veray Public Schools, Indiana.*

Having carefully examined a number of text-books on natural philosophy, I do not hesitate to express my decided opinion in favor of Peck's Ganot. The matter, style, and illustration eminently adapt the work to the popular wants.

From REV. SAMUEL MCKINNEY, D.D., *Pres't Austin College, Huntsville, Texas.*

It gives me pleasure to commend it to teachers. I have taught some classes with it as our text, and must say. for simplicity of style and clearness of illustration, I have found nothing as yet published of equal value to the teacher and pupil.

From C. V. SPEAR, *Principal Maplewood Institute, Pittsfield. Mass.*

I am much pleased with its ample illustrations by plates, and its clearness and simplicity of statement. It covers the ground usually gone over by our higher classes, and contains many fresh illustrations from life or daily occurrences, and new applications of scientific principles to such.

From J. A. BANFIELD, *Superintendent Marshall Public Schools, Michigan.*

I have used Peck's Ganot since 1868, and with increasing pleasure and satisfaction each term. I consider it superior to any other work on physics in its adaptation to our high schools and academies. Its illustrations are superb—better than three times their number of pages of fine print.

From A. SCHUYLER, *Prof. of Mathematics in Baldwin University, Berea, Ohio.*

After a careful examination of Peck's Ganot's Natural Philosophy, and an actual test of its merits as a text-book, I can heartily recommend it as admirably adapted to meet the wants of the grade of students for which it is intended. Its diagrams and illustrations are *unrivaled*. We use it in the Baldwin University.

From D. C. VAN NORMAN, *Principal Van Norman Institute, New York.*

The Natural Philosophy of M. Ganot. edited by Prof. Peck, is, in my opinion, the best work of its kind. for the use intended. ever published in this country. Whether regarded in relation to the natural order of the topics, the precision and clearness of its definitions. or the fullness and beauty of its illustrations, it is certainly, I think, an advance.

☞ For many similar testimonials, see current numbers of the Illustrated Educational Bulletin.

NATURAL SCIENCE—Continued.

CHEMISTRY.

Porter's First Book of Chemistry, · · · · $1 00
Porter's Principles of Chemistry, · · · · · 2 00

The above are widely known as the productions of one of the most eminent scientific men of America. The extreme simplicity in the method of presenting the science, while exhaustively treated, has excited universal commendation.

Darby's Text-Book of Chemistry, · · · · · 1 75

Purely a Chemistry, divesting the subject of matters comparatively foreign to it (such as heat, light, electricity, etc.), but usually allowed to engross too much attention in ordinary school-books.

Gregory's Organic Chemistry, · · · · · · 2 50
Gregory's Inorganic Chemistry, · · · · · 2 50

The science exhaustively treated. For colleges and medical students.

Steele's Fourteen Weeks Course, · · · · · 1 50

A successful effort to reduce the study to the limits of a *single term*, thereby making feasible its general introduction in institutions of every character. The author's felicity of style and success in making the science pre-eminently *interesting* are peculiarly noticeable features. (See page 34.)

Steele's Chemical Apparatus, · · · · · · · *20 00

Adequate to the performance of all the important experiments.

BOTANY.

Thinker's First Lessons in Botany, · · · · 40

For children. The technical terms are largely dispensed with in favor of an easy and familiar style adapted to the smallest learner.

Wood's Object-Lessons in Botany, · · · · 1 50
Wood's American Botanist and Florist, · · 2 50
Wood's New Class-Book of Botany, · · · · 3 50

The standard text-books of the United States in this department. In style they are simple, popular, and lively; in arrangement, easy and natural; in description, graphic and strictly exact. The Tables for Analysis are reduced to a perfect system. More are annually sold than of all others combined.

Wood's Plant Record, · · · · · · · · · · *75

A simple form of Blanks for recording observations in the field.

Wood's Botanical Apparatus, · · · · · · · *8 00

A portable Trunk, containing Drying Press, Knife, Trowel, Microscope, and Tweezers. and a copy of Wood's Plant Record—composing a complete outfit for the collector.

Young's Familiar Lessons, · · · · · · · 2 00
Darby's Southern Botany, · · · · · · · · 2 00

Embracing general Structural and Physiological Botany, with vegetable products, and descriptions of Southern plants, and a complete Flora of the Southern States.

WOOD'S BOTANIES.

TESTIMONIALS.

From PRES. R. B. BURLESON, *Waco University, Texas.*
Wood's Botanies—books that meet every want in their line.

From PRIN. J. G. RALSTON, *Norristown Seminary, Pa.*
We find the "Class-Book" entirely satisfactory.

From PRES. D. F. BITTLE, *Roanoke College, Va.*
Your text-books on Botany are the best for students.

From PROF. W. C. PIERCE, *Baldwin University, Ohio.*
I think his Flora the best we have. His method of analysis is excellent.

From PROF. BLAKESLEE, *State Normal School, Potsdam, N.Y.*
It is admirably concise, yet it does not seem to be deficient or obscure. In paper, print, and binding, the book leaves little to be desired.

From PRES. J. M. GREGORY, *State Agricultural College, Ill.*
I find myself greatly pleased with the perspicuity, compactness, and completeness of the book (Wood's Botanist and Florist). I shall recommend it freely to my friends.

From PROF. A. WINCHELL, *University of Michigan.*
I am free to say that I had been deeply impressed, I may say almost astonished, at the evidences which the work bears of skillful and experienced authorship in this field, and nice and constant adaptation to the wants and conveniences of students of Botany. I pronounce it emphatically an admirable text-book.

From PROF. RICHARD OWEN, *University of Indiana.*
I am well pleased with the evidence of philosophical method exhibited in the general arrangement, as well as with the clearness of the explanations, the ready intelligibility of the analytical tables, and the illustrative aid furnished by the numerous and excellent wood-cuts. I design using the work as a text-book with my next class.

From PRIN. B. R. ANDERSON, *Columbus Union School, Wisconsin.*
I have examined several works with a view to recommending some good text-book on Botany, but I lay them all aside for "Wood's Botanist and Florist." The arrangement of the book is in my opinion excellent, its style fascinating and attractive, its treatment of the various departments of the science is thorough, and last, but far from unimportant, I like the topical form of the questions to each chapter. It seems to embrace the entire science. In fact, I consider it a complete, attractive, and exhaustive work.

From M. A. MARSHALL, *New Haven High School, Conn.*
It has all the excellencies of the well-known Class-Book of Botany by the same author in a smaller book. By a judicious system of condensation, the size of the Flora is reduced one-half, while no species are omitted, and many new ones are added. The descriptions of species are very brief, yet sufficient to identify the plant, and, when taken in connection with the generic description, form a complete description of the plant. The book as a whole will suit the wants of classes better than anything I have yet seen. The adoption of the Botanist and Florist would not require the exclusion of the Class-Book of Botany, as they are so arranged that both might be used by the same class.

From PROF. G. H. PERKINS, *University of Vermont and State Agricultural College.*
I can truly say that the more I examine Wood's Class-Book, the better pleased I am with it. In its illustrations, especially of particulars not easily observed by the student, and the clearness and compactness of its statements, as well as in the territory its flora embraces, it appears to me to surpass any other work I know of. The whole science, so far as it can be taught in a college course, is well presented, and rendered unusually easy of comprehension. The mode of analysis is excellent, avoiding as it does to a great extent those microscopic characters which puzzle the beginner, and using those that are obvious as far as possible. I regard the work as a most admirable one, and shall adopt it as a text-book another year.

PHYSIOLOGY.

Jarvis' Elements of Physiology, · · · · .$ 75
Jarvis' Physiology and Laws of Health, · 1 65

The only books extant which approach this subject with a proper view of the true object of teaching Physiology in schools, viz., that scholars may know how to take care of their own health. In bold contrast with the abstract *Anatomies*, which children learn as they would Greek or Latin (and forget as soon), to *discipline the mind*, are these text-books, using the *science* as a secondary consideration, and only so far as is necessary for the comprehension of the *laws of health*.

Hamilton's Vegetable & Animal Physiology, 1 25

The two branches of the science combined in one volume lead the student to a proper comprehension of the Analogies of Nature.

Steele's Fourteen Weeks Course (see p. 34), · 1 50

ASTRONOMY.

Steele's Fourteen Weeks' Course, · · · · · 1 50

Reduced to a single term, and better adapted to school use than any work heretofore published. Not written for the information of scientific men, but for the inspiration of youth, the pages are not burdened with a multitude of figures which no memory could possibly retain. The whole subject is presented in a clear and concise form. (See p. 34.)

Willard's School Astronomy, · · · · · · 1 00

By means of clear and attractive illustrations, addressing the eye in many cases by analogies, careful definitions of all necessary technical terms, a careful avoidance of verbiage and unimportant matter, particular attention to analysis, and a general adoption of the simplest methods, Mrs. Willard has made the best and most attractive *elementary* Astronomy extant.

McIntyre's Astronomy and the Globes, · · 1 50

A complete treatise for intermediate classes. Highly approved.

Bartlett's Spherical Astronomy, · · · · · 5 00

The West Point course, for advanced classes, with applications to the current wants of Navigation, Geography, and Chronology.

NATURAL HISTORY.

Carll's Child's Book of Natural History, · · 0 50

Illustrating the Animal, Vegetable, and Mineral Kingdoms, with application to the Arts. For beginners. Beautifully and copiously illustrated.

ZOOLOGY.

Chambers' Elements of Zoology, · · · · · 1 50

A complete and comprehensive system of Zoology, adapted for academic instruction, presenting a systematic view of the Animal Kingdom as a portion of external Nature.

Jarvis' Physiology and Laws of Health.

TESTIMONIALS.

From Samuel B. McLane, *Superintendent Public Schools, Keokuk, Iowa.*

I am glad to see a really good text-book on this much neglected branch. This is clear, concise, accurate, and eminently adapted to the *class-room.*

From William F. Wyers, *Principal of Academy, West Chester, Pennsylvania.*

A thorough examination has satisfied me of its superior claims as a text-book to the attention of teacher and taught. I shall introduce it at once.

From H. R. Sanford, *Principal of East Genesee Conference Seminary, N. Y.*

"Jarvis' Physiology" is received, and fully met our expectations. We immediately adopted it.

From Isaac T. Goodnow, *State Superintendent of Kansas—published in connection with the "School Law."*

"Jarvis' Physiology," a common-sense, practical work, with just enough of anatomy to understand the physiological portions. The last six pages, on Man's Responsibility for his own health, are worth the price of the book.

From D. W. Stevens, *Superintendent Public Schools, Fall River, Mass.*

I have examined Jarvis' "Physiology and Laws of Health," which you had the kindness to send to me a short time ago. In my judgment it is far the best work of the kind within my knowledge. It has been adopted as a text-book in our public schools.

From Henry G. Denny, *Chairman Book Committee, Boston, Mass.*

The very excellent "Physiology" of Dr. Jarvis I had introduced into our High School, where the study had been temporarily dropped, believing it to be by far the best work of the kind that had come under my observation; indeed, the reintroduction of the study was delayed for some months, because Dr. Jarvis' book could not be had, and we were unwilling to take any other.

From Prof. A. P. Peabody, D.D., LL.D., *Harvard University.*

* * I have been in the habit of examining school-books with great care, and I hesitate not to say that, of all the text-books on Physiology which have been given to the public, Dr. Jarvis' deserves the first place on the score of accuracy, thoroughness, method, simplicity of statement, and constant reference to topics of practical interest and utility.

From James N. Townsend, *Superintendent Public Schools, Hudson, N. Y.*

Every human being is appointed to take charge of his own body; and of all books written upon this subject, I know of none which will so well prepare one to do this as "Jarvis' Physiology"—that is, in so small a compass of matter. It considers the pure, simple *laws of health* paramount to science: and though the work is thoroughly scientific, it is divested of all cumbrous technicalities, and presents the subject of physical life in a manner and style really charming. It is unquestionably the best text-book on physiology I have ever seen. It is giving great satisfaction in the schools of this city, where it has been adopted as the standard.

From L. J. Sanford, M.D., *Prof. Anatomy and Physiology in Yale College*

Books on human physiology, designed for the use of schools, are more generally a failure perhaps than are school-books on most other subjects.

The great want in this department is met, we think, in the well-written treatise of Dr. Jarvis, entitled "Physiology and Laws of Health." * * The work is not too detailed nor too expansive in any department, and is clear and concise in all. It is not burdened with an excess of anatomical description, nor rendered discursive by many zoological references. Anatomical statements are made to the extent of qualifying the student to attend, understandingly, to an exposition of those functional processes which, collectively, make up health; thus the laws of health are enunciated, and many suggestions are given which, if heeded, will tend to its preservation.

☞ For further testimony of similar character, see current numbers of the Illustrated Educational Bulletin.

NATURAL SCIENCE.

"FOURTEEN WEEKS" IN EACH BRANCH.

By J. DORMAN STEELE, A. M.

Steele's 14 Weeks Course in Chemistry NEW ED., $1 50

Steele's 14 Weeks Course in Astronomy · 1 50

Steele's 14 Weeks Course in Philosophy · 1 50

Steele's 14 Weeks Course in Geology. · 1 50

Steele's 14 Weeks Course in Physiology · 1 50

Our Text-Books in these studies are, as a general thing, dull and uninteresting. They contain from 400 to 600 pages of dry facts and unconnected details. They abound in that which the student cannot learn, much less remember. The pupil commences the study, is confused by the fine print and coarse print, and neither knowing exactly what to learn nor what to hasten over, is crowded through the single term generally assigned to each branch, and frequently comes to the close without a definite and exact idea of a single scientific principle.

Steele's Fourteen Weeks Courses contain only that which every well-informed person should know, while all that which concerns only the professional scientist is omitted. The language is clear, simple, and interesting, and the illustrations bring the subject within the range of home life and daily experience. They give such of the general principles and the prominent facts as a pupil can make familiar as household words within a single term. The type is large and open; there is no fine print to annoy; the cuts are copies of genuine experiments or natural phenomena, and are of fine execution.

In fine, by a system of condensation peculiarly his own, the author reduces each branch to the limits of a single term of study, while sacrificing nothing that is essential, and nothing that is usually retained from the study of the larger manuals in common use. Thus the student has rare opportunity to *economize his time*, or rather to employ that which he has to the best advantage.

A notable feature is the author's charming "style," fortified by an enthusiasm over his subject in which the student will not fail to partake. Believing that Natural Science is full of fascination, he has moulded it into a form that attracts the attention and kindles the enthusiasm of the pupil.

The recent editions contain the author's "Practical Questions" on a plan never before attempted in scientific text-books. These are questions as to the nature and cause of common phenomena, and are not directly answered in the text, the design being to test and promote an intelligent use of the student's knowledge of the foregoing principles.

Steele's General Key to his Works · · · · *1 50

This work is mainly composed of Answers to the Practical Questions and Solutions of the Problems in the author's celebrated "Fourteen Weeks Courses" in the several sciences, with many hints to teachers, minor Tables, &c. Should be on every teacher's desk.

Steele's 14 Weeks in each Science.

TESTIMONIALS.

From L. A. BIKLE, President N. C. College.

I have not been disappointed. Shall take pleasure in introducing this series.

From J. F. COX, Prest. Southern Female College, Ga.

I am much pleased with these books, and expect to introduce them.

From J. R. BRANHAM, Prin. Brownsville Female College, Tenn.

They are capital little books, and are now in use in our institution.

From W. H. GOODALE, Professor Readville Seminary, La.

We are using your 14 Weeks Course, and are much pleased with them.

From W. A. BOLES, Supt. Shelbyville Graded School, Ind.

They are as entertaining as a story book, and much more improving to the mind.

From S. A. SNOW, Principal of High School, Uxbridge, Mass.

Steele's 14 Weeks Courses in the Sciences are a perfect success.

From JOHN W. DOUGHTY, Newburg Free Academy, N. Y.

I was prepared to find Prof. Steele's Course both attractive and instructive. My highest expectations have been fully realized.

From J. S. BLACKWELL, Prest. Ghent College, Ky.

Prof. Steele's unexampled success in providing for the wants of academic classes, has led me to look forward with high anticipations to his forthcoming issue.

From J. F. COOK, Prest. La Grange College, Mo.

I am pleased with the neatness of these books and the delightful diction. I have been teaching for years, and have never seen a lovelier little volume than the Astronomy.

From M. W. SMITH, Prin. of High School, Morrison, Ill.

They seem to me to be admirably adapted to the wants of a public school, containing, as they do, a sufficiently comprehensive arrangement of elementary principles to excite a healthy thirst for a more thorough knowledge of those sciences.

From J. D. BARTLEY, Prin. of High School, Concord, N. H.

They are just such books as I have looked for, viz., those of interesting style, not cumbersome and filled up with things to be omitted by the pupil, and yet sufficiently full of facts for the purpose of most scholars in these sciences in our high schools ; there is nothing but what a pupil of average ability can thoroughly master.

From ALONZO NORTON LEWIS, Principal of Parker Academy, Conn.

I consider Steele's Fourteen Weeks Courses in Philosophy, Chemistry, &c., the *best* school-books that have been issued in this country.

As an introduction to the various branches of which they treat, and especially . for that numerous class of pupils who have not the time for a more extended course, I consider them *invaluable.*

From EDWARD BROOKS, Prin. State Normal School, Millersville, Pa.

At the meeting of Normal School Principals, I presented the following resolution, which was unanimously adopted : "*Resolved*, That Steele's 14 Weeks Courses in Natural Philosophy and Astronomy, or an amount equivalent to what is contained in them, be adopted for use in the State Normal Schools of Pennsylvania." The works themselves will be adopted by at least three of the schools, and, I presume, by them all.

LITERATURE.

Cleveland's Compendiums · · · · each, $*2 50

ENGLISH LITERATURE. AMERICAN LITERATURE.
ENGLISH LITERATURE OF THE XIXTH CENTURY.

In these volumes are gathered the cream of the literature of the English speak,
ing people for the school-room and the general reader. Their reputation is
national. More than 125,000 copies have been sold.

Boyd's English Classics · · · · · each, *1 25

MILTON'S PARADISE LOST. THOMSON'S SEASONS.
YOUNG'S NIGHT THOUGHTS. POLLOK'S COURSE OF TIME.
COWPER'S TASK, TABLE TALK, &c. LORD BACON'S ESSAYS.

This series of annotated editions of great English writers, in prose and poetry,
is designed for critical reading and parsing in schools. Prof. J. R. Boyd proves
himself an editor of high capacity, and the works themselves need no encomium.
As auxiliary to the study of Belles Lettres, etc., these works have no equal.

Pope's Essay on Man · · · · · · · · · *20

Pope's Homer's Iliad · · · · · · · · · *80

The metrical translation of the great poet of antiquity, and the matchless
" Essay on the Nature and State of Man," by ALEXANDER POPE, afford superior
exercise in literature and parsing.

AESTHETICS.

Huntington's Manual of the Fine Arts · · *1 75

A view of the rise and progress of Art in different countries, a brief
account of the most eminent masters of Art, and an analysis of the principles of Art. It is complete in itself, or may precede to advantage the
critical work of Lord Kames.

Boyd's Kames' Elements of Criticism · · *1 75

The best edition of this standard work; without the study of which
none may be considered proficient in the science of the Perceptions. No
other study can be pursued with so marked an effect upon the taste and
refinement of the pupil.

POLITICAL ECONOMY.

Champlin's Lessons on Political Economy 1 25

An improvement on previous treatises, being shorter, yet containing
every thing essential, with a view of recent questions in finance, etc.,
which is not elsewhere found.

CLEVELAND'S COMPENDIUMS.

TESTIMONIALS.

From the New Englander.
This is the very best book of the kind we have ever examined.

From GEORGE B. EMERSON, Esq., *Boston.*
The Biographical Sketches are just and discriminating; the selections are admir-able, and I have adopted the work as a text-book for my first class.

From PROF. MOSES COIT TYLER, *of the Michigan University.*
I have given your book a thorough examination, and am greatly delighted with it; and shall have great pleasure in directing the attention of my classes to a work which affords so admirable a bird's-eye view of recent "English Literature."

From the Saturday Review.
It acquaints the reader with the characteristic method, tone, and quality of all the chief notabilities of the period, and will give the careful student a better idea of the recent history of English Literature than nine educated Englishmen in ten possess.

From the Methodist Quarterly Review, New York.
This work is a transcript of the best American mind; a vehicle of the noblest American spirit. No parent who would introduce his child to a knowledge of our country's literature, and at the same time indoctrinate his heart in the purest prin-ciples, need fear to put this manual in the youthful hand.

From REV. C. PEIRCE, *Principal, West Newton, Mass.*
I do not believe the work is to be found from which, within the same limits. so much interesting and valuable information in regard to English writers and English literature of every age, can be obtained; and it deserves to find a place in all our high schools and academies, as well as in every private library.

From the Independent.
The work of selection and compilation—requiring a perfect familiarity with the whole range of English literature, a judgment clear and impartial, a taste at once delicate and severe, and a most sensitive regard to purity of thought or feeling—has been better accomplished in this than in any kindred volume with which we are acquainted.

From the Christian Examiner.
To form such a Compendium, good taste, fine scholarship familiar acquaintance with English literature, unwearied industry, tact acquired by practice, an interest in the culture of the young, a regard for truth, purity, philanthropy, religion, as the highest attainment and the highest beauty,—all these were needed, and they are united in Mr. Cleveland.

CHAMPLIN'S POLITICAL ECONOMY.

From J. L. BOTHWELL, *Prin. Public School No. 14, Albany, N.Y.*
I have examined Champlin's Political Economy with much pleasure, and shall be pleased to put it into the hands of my pupils. In quantity and quality I think it superior to anything that I have examined.

From PRES. N. E. COBLEIGH, *East Tennessee Wesleyan University.*
An examination of Champlin's Political Economy has satisfied me that it is the book I want. For brevity and compactness, division of the subject, and clear state-ment, and for appropriateness of treatment, I consider it a better text-book than any other in the market.

From the Evening Mail, New York.
A new interest has been imparted to the science of political economy since we have been necessitated to raise such vast sums of money for the support of the gov-ernment. The time, therefore, is favorable for the introduction of works like the above. This little volume of two hundred pages is intended for beginners, for the common school and academy. It is intended as a basis upon which to rear a more elaborate superstructure. There is nothing in the principles of political economy above the comprehension of average scholars, when they are clearly set forth. This seems to have been done by President Champlin in an easy and graceful manner.

ELOCUTION.

Taverner Graham's Reasonable Elocution, $1 25
Based upon the belief that true Elocution is the right interpretation of THOUGHT, and guiding the student to an intelligent appreciation, instead of a merely mechanical knowledge, of its rules.

Zachos' Analytic Elocution 1 50
All departments of elocution—such as the analysis of the voice and the
sentence, phonology, rhythm, expression, gesture, &c.—are here arranged
for instruction in classes, illustrated by copious examples.

Sherwood's Self Culture 1 00
Self-culture in reading, speaking, and conversation—a very valuable
treatise to those who would perfect themselves in these accomplishments.

SPEAKERS.

Northend's Little Orator, *60—Child's Speaker*60
Two little works of the same grade but different selections, containing
simple and attractive pieces for children under twelve years of age.

Northend's Young Declaimer *75
Northend's National Orator *1 25
Two volumes of Prose, Poetry, and Dialogue, adapted to intermediate and grammar classes respectively.

Northend's Entertaining Dialogues *1 25
Extracts eminently adapted to cultivate the dramatic faculties, as well
as entertain an audience.

Swett's Common School Speaker *1 25
Selections from recent literature.

Raymond's Patriotic Speaker *2 00
A superb compilation of modern eloquence and poetry, with original
dramatic exercises. Nearly every eminent *living* orator is represented,
without distinction of place or party.

COMPOSITION, &c.

Brookfield's First Book in Composition . 50
Making the cultivation of this important art feasible for the smallest
child. By a new method, to induce and stimulate thought.

Boyd's Composition and Rhetoric 1 50
This work furnishes all the aid that is needful or can be desired in
the various departments and styles of composition, both in prose and verse.

Day's Art of Rhetoric 1 25
Noted for exactness of definition, clear limitation, and philosophical
development of subject; the large share of attention given to Invention,
as a branch of Rhetoric, and the unequalled analysis of style

MENTAL PHILOSOPHY.

Mahan's Intellectual Philosophy $1 75

The subject exhaustively considered. The author has evinced learning, candor, and independent thinking.

Mahan's Science of Logic 2 00

A profound analysis of the laws of thought. The system possesses the merit of being intelligible and self consistent. In addition to the author's carefully elaborated views, it embraces results attained by the ablest minds of Great Britain, Germany, and France, in this department.

Boyd's Elements of Logic 1 25

A systematic and philosophic condensation of the subject, fortified with additions from Watts, Abercrombie, Whately, &c.

Watts on the Mind 50

The Improvement of the Mind, by Isaac Watts, is designed as a guide for the attainment of useful knowledge. As a text-book it is unparalleled; and the discipline it affords cannot be too highly esteemed by the educator.

MORALS.

Peabody's Moral Philosophy 1 25

A shot course; by the Professor of Christian Morals, Harvard University—for the Freshman Class and for High Schools.

Alden's Text-Book of Ethics 60

For young pupils. To aid in systematizing the ethical teachings of the Bible, and point out the coincidences between the instructions of the sacred volume and the sound conclusions of reason.

Willard's Morals for the Young 75

Lessons in conversational style to inculcate the elements of moral philosophy. The study is made attractive by narratives and engravings.

GOVERNMENT.

Howe's Young Citizen's Catechism 75

Explaining the duties of District, Town, City, County, State, and United States Officers, with rules for parliamentary and commercial business—that which every future " sovereign " ought to know, and so few are taught.

Young's Lessons in Civil Government . . 1 25

A comprehensive view of Government, and abstract of the laws showing the rights, duties, and responsibilities of citizens.

Mansfield's Political Manual 1 25

This is a complete view of the theory and practice of the General and State Governments of the United States, designed as a text-book. The author is an esteemed and able professor of constitutional law, widely known for his sagacious utterances in matters of statecraft through the public press. Recent events teach with emphasis the vital necessity that the rising generation should comprehend the noble polity of the American government, that they may act intelligently when endowed with a voice in it.

MODERN LANGUAGE.

French and English Primer, ...$	10
German and English Primer, ...	10
Spanish and English Primer, ...	10

The names of common objects properly illustrated and arranged in easy lessons.

Ledru's French Fables, ...	75
Ledru's French Grammar, ... 1	00
Ledru's French Reader, . . . φ 1	00

The author's long experience has enabled him to present the most thoroughly practical text-books extant, in this branch. The system of pronunciation (by phonetic illustration) is original with this author, and will commend itself to all American teachers, as it enables their pupils to secure an absolutely correct pronunciation without the assistance of a native master. This feature is peculiarly valuable also to "self-taught" students. The directions for ascertaining the gender of French nouns—also a great stumbling-block—are peculiar to this work, and will be found remarkably competent to the end proposed. The criticism of teachers and the test of the school-room is invited to this excellent series, with confidence.

Worman's French Echo, ... 1	25

To teach conversational French by actual practice, on an entirely new plan, which recognizes the importance of the student learning *to think* in the language which he speaks. It furnishes an extensive vocabulary of words and expressions in common use, and suffices to free the learner from the embarrassments which the peculiarities of his own tongue are likely to be to him, and to make him thoroughly familiar with the use of proper idioms.

Worman's German Echo, ... 1	25

On the same plan. See Worman's German Series, page 42.

Pujol's Complete French Class-Book, . . . 2	25

Offers, in one volume, methodically arranged, a complete French course —usually embraced in series of from five to twelve books, including the bulky and expensive Lexicon. Here are Grammar, Conversation, and choice Literature—selected from the best French authors. Each branch is thoroughly handled ; and the student, having diligently completed the course as prescribed, may consider himself, without further application, *au fait* in the most polite and elegant language of modern times.

Maurice-Poitevin's Grammaire Francaise, . 1	00

American schools are at last supplied with an American edition of this famous text-book. Many of our best institutions have for years been procuring it from abroad rather than forego the advantages it offers. The policy of putting students who have acquired some proficiency from the ordinary text-books, into a Grammar written in the vernacular, can not be too highly commended. It affords an opportunity for finish and review at once ; while embodying abundant practice of its own rules.

Joynes' French Pronunciation, ...	30
Willard's Historia de los Estados Unidos, . 2	00

The History of the United States, translated by Professors Tolon and De Tornos, will be found a valuable, instructive, and entertaining reading-book for Spanish classes.

Pujol's Complete French Class-Book.

TESTIMONIALS.

From PROF. ELIAS PEISSNER, *Union College.*

I take great pleasure in recommending Pujol and Van Norman's French Class-Book, as there is no French grammar or class-book which can be compared with it in completeness, system, clearness, and general utility.

From EDWARD NORTH, *President of Hamilton College.*

I have carefully examined Pujol and Van Norman's French Class-Book, and am satisfied of its superiority, for college purposes, over any other heretofore used. We shall not fail to use it with our next class in French.

From A. CURTIS, *Pres't of Cincinnati Literary and Scientific Institute.*

I am confident that it may be made an instrument in conveying to the student, in from six months to a year, the art of speaking and writing the French with almost native fluency and propriety.

From HIRAM ORCUTT, A. M., *Prin. Glenwood and Tilden Ladies' Seminaries.*

I have used Pujol's French Grammar in my two seminaries, exclusively, for more than a year, and have no hesitation in saying that I regard it the best text-book in this department extant. And my opinion is confirmed by the testimony of Prof. F. De Launay and Mademoiselle Marindin. They assure me that the book is eminently accurate and practical, as tested in the school-room.

From PROF. THEO. F. DE FUMAT, *Hebrew Educational Institute, Memphis, Tenn.*

M. Pujol's French Grammar is one of the best and most practical works. The French language is chosen and elegant in style—modern and easy. It is far superior to the other French class-books in this country. The selection of the conversational part is very good, and will interest pupils; and being all completed in only one volume, it is especially desirable to have it introduced in our schools.

From PROF. JAMES H. WORMAN, *Bordentown Female College, N. J.*

The work is upon the same plan as the text-books for the study of French and English published in Berlin, for the study of those who have not the aid of a teacher, and these books are considered, by the first authorities, the best books. In most of our institutions, Americans teach the modern languages, and heretofore the trouble has been to give them a text-book that would dispose of the difficulties of the French pronunciation. This difficulty is successfully removed by P. and Van N., and I have every reason to believe it will soon make its way into most of our best schools.

From PROF. CHARLES S. DOD, *Ann Smith Academy, Lexington, Va.*

I cannot do better than to recommend "Pujol and Van Norman." For comprehensive and systematic arrangement, progressive and thorough development of all grammatical principles and idioms, with a due admixture of theoretical knowledge and practical exercise, I regard it as superior to any (other) book of the kind.

From A. A. FORSTER, *Prin. Pinehurst School, Toronto, C. W.*

I have great satisfaction in bearing testimony to M. Pujol's System of French Instruction, as given in his complete class-book. For clearness and comprehensiveness, adapted for all classes of pupils, I have found it superior to any other work of the kind, and have now used it for some years in my establishment with great success.

From PROF. OTTO FEDDER, *Maplewood Institute, Pittsfield, Mass.*

The conversational exercises will prove an immense saving of the hardest kind of labor to teachers. There is scarcely any thing more trying in the way of teaching language, than to rack your brain for short and easily intelligible bits of conversation, and to repeat them time and again with no better result than extorting at long intervals a doubting "oui," or a hesitating "non, monsieur"

☞ For further testimony of a similar character, see special circular, and current numbers of the Educational Bulletin.

GERMAN.

A COMPLETE COURSE IN THE GERMAN,
By JAMES H. WORMAN, A. M.

Worman's Elementary German Grammar . $1 50
Worman's Complete German Grammar . 2 00

These volumes are designed for intermediate and advanced classes respectively. Though following the same general method with "Otto" (that of 'Gaspey'). our author differs essentially in its application. He is more practical, more systematic, more accurate, and besides introduces a number of invaluable features which have never before been combined in a German grammar.

Among other things, it may be claimed for Prof. Worman that he has been *the first* to introduce in an American text-book for learning German, a system of analogy and comparison with other languages. Our best teachers are also enthusiastic about his methods of inculcating the art of speaking, of understanding the spoken language, of correct pronunciation; the sensible and convenient original classification of nouns (in four declensions), and of irregular verbs, also deserves much praise. We also note the use of heavy type to indicate etymological changes in the paradigms, and, in the exercises, the parts which specially illustrate preceding rules.

Worman's Elementary German Reader . . 1 25
Worman's Collegiate German Reader . . . 2 00

The finest and most judicious compilation of classical and standard German Literature. These works embrace, progressively arranged, selections from the masterpieces of Goethe, Schiller, Korner, Seume, Uhland, Freiligrath, Heine, Schlegel, Holty, Lenau, Wieland, Herder, Lessing, Kant, Fichte, Schelling, Winkelmann, Humboldt, Ranke, Raumer, Menzel, Gervinus, &c., and contains complete Goethe's "Iphigenie," Schiller's "Jungfrau;" also, for instruction in modern conversational German, Benedix's "Eigensinn."

There are besides, Biographical Sketches of each author contributing, Notes, explanatory and philological (after the text), Grammatical References to all leading grammars, as well as the editor's own, and an adequate Vocabulary.

Worman's German Echo 1 25

Consists of exercises in colloquial style entirely in the German, with an adequate vocabulary, not only of words but of idioms. The object of the system developed in this work (and its companion volume in the French) is to break up the laborious and tedious habit of *translating the thoughts*, which is the student's most effectual bar to fluent conversation, and to lead him to *think in the language in which he speaks*. As the exercises illustrate scenes in actual life, a considerable knowledge of the manners and customs of the German people is also acquired from the use of this manual.

Worman's German Copy-Books, 3 Numbers, each   15

On the same plan as the most approved systems for English penmanship, with progressive copies.

42

Worman's German Grammars.

TESTIMONIALS.

From Prof. R. W. Jones, Petersburg Female College, Va.

From what I have seen of the work it is almost certain *I shall introduce it* into this institution.

From Prof. G. Campbell, University of Minnesota.

A valuable addition to our school-books, and will find many friends, and do great good.

From Prof. O. H P. Corprew, Mary Military Inst, Md.

I am better pleased with them than any I have ever taught. I have already ordered through our booksellers.

From Prof. R. S. Kendall, Vernon Academy, Conn.

I at once put the Elementary Grammar into the hands of a class of beginners, and have used it *with great satisfaction.*

From Prof. D. E. Holmes, Berlin Academy, Wis.

Worman's German works are *superior.* I shall use them hereafter in my German classes.

From Prof. Magnus Buchholtz, Hiram College, Ohio.

I have examined the Complete Grammar, and find it *excellent.* You may rely that it will be used here.

From Prin. Thos. W. Tobey, Paducah Female Seminary, Ky.

The Complete German Grammar is worthy of an extensive circulation. It is *admirably adapted* to the class-room. I shall use it.

From Prof. Alex. Rosenspitz, Houston Academy, Texas.

Bearer will take and pay for 3 dozen copies. Mr. Worman deserves the approbation and esteem of the teacher and the thanks of the student.

From Prof. G. Malmene, Augusta Seminary, Maine.

The Complete Grammar cannot fail to *give great satisfaction* by the simplicity of its arrangement, and by its completeness.

From Prin. Oval Pirkey, Christian University, Mo.

Just such a series as is positively necessary. I do hope the author will succeed as well in the French, &c., as he has in the German.

From Prof. S. D. Hillman, Dickinson College, Pa.

The class have lately commenced, and my examination thus far warrants me in saying that I regard it as *the best* grammar for instruction in the German.

From Prin. Silas Livermore, Bloomfield Seminary, Mo.

I have found a classically and scientifically educated Prussian gentleman whom I propose to make German instructer. I have shown him both your German grammars. He has expressed *his approbation* of them generally.

From Prof. Z. Test, Howland School for Young Ladies, N. Y.

I shall introduce the books. From a cursory examination I have no hesitation in pronouncing the Complete Grammar *a decided improvement* on the text-books at present in use in this country.

From Prof. Lewis Kistler, Northwestern University, Ill.

Having looked through the Complete Grammar with some care I must say that you have produced *a good book;* you may be awarded with this gratification—that your grammar promotes the facility of learning the German language, and of becoming acquainted with its rich literature.

From Pres. J. P. Rous, Stockwell Collegiate Inst., Ind.

I supplied a class with the Elementary Grammar, and it gives *complete satisfaction.* The conversational and reading exercises are well calculated to illustrate the principles, and lead the student on an easy yet thorough course. I think the Complete Grammar equally attractive.

THE CLASSICS.

LATIN.

Silber's Latin Course, ... $1 25

The book contains an Epitome of Latin Grammar, followed by Reading Exercises, with explanatory Notes and copious References to the leading Latin Grammars, and also to the Epitome which precedes the work. Then follow a Latin-English Vocabulary and Exercises in Latin Prose Composition, being thus complete in itself, and a very suitable work to put in the hands of one about to study the language.

Searing's Virgil's Æneid, ... 2 25

It contains only the first six books of the Æneid. 2. A very carefully constructed Dictionary. 3. Sufficiently copious Notes. 4. Grammatical references to four leading Grammars. 5. Numerous Illustrations of the highest order, 6. A superb Map of the Mediterranean and adjacent countries. 7. Dr. S. H. Taylor's "Questions on the Æneid." 8. A Metrical Index, and an Essay on the Poetical Style. 9. A photographic *fac simile* of an early Latin M.S. 10. The text according to Jahn, but paragraphed according to Ladewig. 11. Superior mechanical execution.

Blair's Latin Pronunciation, ... 1 00

An inquiry into the proper sounds of the Language during the Classical Period. By Prof. Blair, of Hampden Sidney College, Va.

Andrews & Stoddard's Latin Grammar, *1 50
Andrews' Questions on the Grammar, . *0 15
Andrews' Latin Exercises, · · · · · · *1 25
Andrews' Viri Romæ, · · · · · · · · *1 25
Andrews' Sallust's Jugurthine War, &c. *1 50
Andrews' Eclogues & Georgics of Virgil, *1 50
Andrews' Cæsar's Commentaries, · · · · *1 50
Andrews' Ovid's Metamorphoses, · · · *1 25

GREEK.

Crosby's Greek Grammar, · · · · · · · 2 00
Crosby's Xenophon's Anabasis, · · · · · 1 25

Searing's Homer's Iliad, · · · · · · · ·

MYTHOLOGY.
Dwight's Grecian and Roman Mythology.

School edition, $1 25;   University edition, *3 00

A knowledge of the fables of antiquity, thus presented in a systematic form, is as indispensable to the student of general literature as to him who would peruse intelligently the classical authors. The mythological allusions so frequent in literature are readily understood with such a Key as this.

SEARING'S VIRGIL.

SPECIMEN FRAGMENTS OF LETTERS.

" I adopt it gladly."—Prin. V. Dabney, *Loudoun School, Va.*

" I like Searing's Virgil."—Prof. Bristol, *Ripon College, Wis.*

" Meets my desires very thoroughly."—Prof. Clark, *Berea College, Ohio.*

" Superior to any other edition of Virgil."—Pres. Hall, *Macon College, Mo.*

" Shall adopt it at once."—Prin. B. P. Baker, *Searcy Female Institute. Ark.*

" Your Virgil is a *beauty*."—Prof. W. H. De Motte, *Illinois Female College.*

" After use, I regard it the best."—Prin. G. H. Barton, *Rome Academy, N. Y.*

" We like it better every day."—Prin. R. K. Buehrle, *Allentown Academy, Pa.*

" I am delighted with your Virgil."—Prin. W. T. Leonard, *Pierce Academy, Mass.*

" Stands well the test of class-room."—Prin. F. A. Chase, *Lyons Col. Inst., Iowa.*

" I do not see how it can be improved."—Prin. N. F. D. Browne, *Charl. Hall, Md.*

" The most complete that I have seen."—Prin. A. Brown, *Columbus High School, Ohio.*

" Our Professor of Language very highly approves."—Supt. J. G. James, *Texas Military Institute.*

" It responds to a want long felt by teachers. It is beautiful and complete."—Prof. Brooks. *University of Minnesota.*

" The ideal edition. We want a few more classics of the same sort."—Prin. C. F. P. Bancroft, *Lookout Mountain Institute. Tenn.*

" I certainly have never seen an edition so complete with important requisites for a student. nor with such fine text and general mechanical execution."—Pres. J. R. Park, *University of Deseret, Utah.*

" It is charming both in its design and execution. And, on the whole, I think it is the best thing of the kind that I have seen."—Prof. J. De F. Richards, *Pres. pro tem. of University of Alabama.*

" In beauty of execution. in judicious notes, and. in an adequate vocabulary, it merits all praise. I shall recommend its introduction."—Pres. J. K. Patterson, *Kentucky Agricultural and Mechanical College.*

" Containing a good vocabulary and judicious notes, it will enable the industrious student to acquire an accurate knowledge of the most interesting part of Virgil's works."—Prof. J. T. Dunklin, *East Alabama College.*

" It wants no element of completeness. It is by far the best classical text-book with which I am acquainted. The notes are just right. They help the student when he most needs help."—Prin. C. A. Bunker, *Caledonia Grammar School, Vt.*

" I have examined Searing's Virgil with interest, and find that it more nearly meets the wants of students than that of any other edition with which I am acquainted. I am able to introduce it to some extent at once."—Prin. J. Easter, *East Genesee Conference Seminary.*

" I have been wishing to get a sight of it, and it exceeds my expectations. It is a beautiful book in every respect, and bears evidence of careful and critical study. The engravings add instruction as well as interest to the work. I shall recommend it to my classes."—Prin. Chas. H. Chandler, *Glenwood Ladies' Seminary.*

" A. S. Barnes & Co. have published an edition of the first six books of Virgil's Æneid. which is superior to its predecessors in several respects. The publishers have done a good service to the cause of classical education, and the book deserves a large circulation."—Prof. George W. Collord, *Brooklyn Polytechnic, N. Y.*

" My attention was called to Searing's Virgil by the fact of its containing a vocabulary which would obviate the necessity of procuring a lexicon. But use in the class-room has impressed me most favorably with the accuracy and just proportion of its notes, and the general excellence of its grammatical suggestions. The general character of the book in its paper, its typography, and its engravings is highly commendable, and the fac-simile manuscript is a valuable feature. I take great pleasure in commending the book to all who do not wish a complete edition of Virgil. It suits our short school courses admirably."—Henry L. Boltwood, *Master of Princeton High School, Ill.*

RECORDS.

Cole's Self-Reporting Class-Book, ... *$0 50

For saving the Teacher's labor in averaging. At each opening are a full set of Tables showing any scholar's standing at a glance and entirely obviating the necessity of computation.

Tracy's School-Record, *0 75.   Pocket edition, *0 65

For keeping a simple but exact record of Attendance, Deportment, and Scholarship. The larger edition contains also a Calendar, an extensive list of Topics for Compositions and Colloquies, Themes for Short Lectures, Suggestions to Young Teachers, etc.

Brooks' Teacher's Register, ... *1 00

Presents at one view a record of Attendance, Recitations, and Deportment for the whole term.

Carter's Record and Roll-Book, ... *1 50

This is the most complete and convenient Record offered to the public. Besides the usual spaces for General Scholarship, Deportment, Attendance, etc., for each name and day, there is a space in red lines enclosing six minor spaces in blue for recording Recitations.

National School Diary, ... Per dozen, *1 00

A little book of blank forms for weekly report of the standing of each scholar, from teacher to parent. A great convenience.

REWARDS.

National School Currency, ... Per set, *$1 50

A little box containing certificates in the form of Money. The most entertaining and stimulating system of school rewards. The scholar is paid for his merits and fined for his shortcomings. Of course the most faithful are the most successful in business. In this way the use and value of money and the method of keeping accounts are also taught. One box of Currency will supply a school of fifty pupils.

TACTICS.

The Boy Soldier, ... 75

Complete Infantry Tactics for Schools, with illustrations, for the use of those who would introduce this pleasing relaxation from the confining duties of the desk.

CHARTS.

McKenzie's Elocutionary Chart, · · · · · $3 50

Baade's Reading Case, · · · · · · · · · *10 00

This remarkable piece of school-room furniture is a receptacle containing a number of primary cards. By an arrangement of slides on the front, one sentence at a time is shown to the class. Twenty-eight thousand transpositions may be made, affording a variety of progressive exercises which no other piece of apparatus offers. One of its best features is, that it is so exceedingly simple as not to get out of order, while it may be operated with one finger.

Marcy's Eureka Tablet, · · · · · · · · · *1 50

A new system for the Alphabet, by which it may be taught without fail in nine lessons.

Scofield's School Tablets, · · · · · · · · *8 00

On Five Cards, exhibiting Ten Surfaces. These Tablets teach Orthography, Reading, Object-Lessons, Color, Form, etc.

Watson's Phonetic Tablets, · · · · · · · *8 00

Four Cards, and Eight Surfaces ; teaching Pronunciation and Elocution phonetically—for class exercises.

Page's Normal Chart, · · · · · · · · · · *3 75

The whole science of Elementary Sounds tabulated. By the author of Page's Theory and Practice of Teaching.

Clark's Grammatical Chart, · · · · · · · *3 75

Exhibits the whole Science of Language in one comprehensive diagram.

Davies' Mathematical Chart, · · · · · · · *75

Mathematics made simple to the eye.

Monteith's Reference Maps (School Series), · ·*20 00

Eight Numbers. Mounted on Rollers. Names all laid down in small type, so that to the pupil at a short distance they are Outline Maps, while they serve as *their own key* to the teacher.

Willard's Chronographers, · · · · · Each, *2 00

Historical. Four Numbers. Ancient Chronographer ; English Chronographer ; American Chronographer ; Temple of Time (general). Dates and Events represented to the eye.

APPARATUS.

Harrington's Geometrical Blocks, · · · ·*$10 00

These patented blocks are *hinged*, so that each form can be dissected.

Harrington's Fractional Blocks, · · · · · *8 00

Steele's Chemical Apparatus, · ·*20 00

Steele's Philosophical Apparatus, (see p.28) *125 00

Steele's Geological Cabinet, (see p.28) · · ·*40 00

Wood's Botanical Apparatus, (see p.30) · · *8 00

Bock's Physiological Apparatus, · · · · .175 00

MUSIC.

Jepson's Music Readers. 3 vols. . . . Each, 75 cts,

These are not books from which children simply learn songs, par-rot-like, but teach the subject progressively—the scholar learning to read music by methods similar to those employed in teaching him to read printed language. Any teacher, however ignorant of music, pro-vided he can, upon trial, simply sound the scale, may teach it without assistance, and will end by being a good singer himself. The "Ele-mentary Music Reader," or first volume, heretofore issued by another publisher, has attained results in the State of Connecticut, where only it has been known, entirely unprecedented in the history of teaching music. The two companion volumes carry the same method into the higher grades.

Nash & Bristow's Cantara. No. 1, $1.15; No. 2, $1.40

The first volume is a complete musical text-book for schools of every grade. No. 2 is a choice selection of Solos and Part Songs. The authors are Directors of Music in the public schools of New York City, in which these books are the standard of instruction.

Curtis' Little Singer, ... $0 60

Curtis' School Vocalist, ... 1 00

Kingsley's School-Room Choir, ... 60

Kingsley's Young Ladies' Harp, ... 1 00

Hager's Echo, ... 75

Perkins' Sabbath Carols (for Sunday-Schools), . . 35

Phillips' Singing Annual do. do. . 25

DEVOTION.

Brooks' School Manual of Devotion, . . . $0 75

This volume contains daily devotional exercises, consisting of a hymn, selections of Scripture for alternate reading by teacher and pupils, and a prayer. Its value for opening and closing school is apparent.

Brooks' School Harmonist, ... 75

Contains appropriate *tunes* for each hymn in the "Manual of Devo-tion" described above.

THE
TEACHERS' LIBRARY.

Object Lessons—Welch *$1 00

This is a complete exposition of the popular modern system of
"object teaching," for teachers of primary classes.

Theory and Practice of Teaching—Page . . *1 50

This volume has, without doubt, been read by two hundred thousand
teachers, and its popularity remains undiminished—large editions
being exhausted yearly. It was the pioneer, as it is now the patri-
arch of professional works for teachers.

The Graded School—Wells *1 25

The proper way to organize graded schools is here illustrated. The
author has availed himself of the best elements of the several systems
prevalent in Boston, New York, Philadelphia, Cincinnati, St. Louis,
and other cities.

The Normal—Holbrook *1 50

Carries a working school on its visit to teachers, showing the most
approved methods of teaching all the common branches, including the
technicalities, explanations, demonstrations, and definitions intro-
ductory and peculiar to each branch.

The Teachers' Institute—Fowle *1 25

This is a volume of suggestions inspired by the author's experience
at institutes, in the instruction of young teachers. A thousand points
of interest to this class are most satisfactorily dealt with.

Schools and Schoolmasters—Dickens . . . *1 25

Appropriate selections from the writings of the great novelist.

The Metric System—Davies *1 50

Considered with reference to its general introduction, and embracing the views of John Quincy Adams and Sir John Herschel.

The Student '—The Educator—Phelps . each,*1 50
The Discipline of Life—Phelps *1 75

The authoress of these works is one of the most distinguished
writers on education; and they cannot fail to prove a valuable addition to the School and Teachers' Libraries, being in a high degree
both interesting and instructive.

A Scientific Basis of Education—Hecker . . *2 50

Adaptation of study and classification by temperaments.

American Education—Mansfield $1 50

A treatise on the principles and elements of education, as practiced in this country, with ideas towards distinctive republican and Christian education.

American Institutions—De Tocqueville . .*1 50

A valuable index to the genius of our Government.

Universal Education—Mayhew *1 75

The subject is approached with the clear, keen perception of one who has observed its necessity, and realized its feasibility and expediency alike. The redeeming and elevating power of improved common schools constitutes the inspiration of the volume.

Higher Christian Education—Dwight . . .*1 50

A treatise on the principles and spirit, the modes, directions, and results of all true teaching; showing that right education should appeal to every element of enthusiasm in the teacher'.. ature.

Oral Training Lessons—Barnard *1 00

The object of this very useful work is to furnish material for instructors to impart orally to their classes, in branches not usually taught in common schools, embracing all departments of Natural Science and much general knowledge.

Lectures on Natural History—Chadbourne * 75

Affording many themes for oral instruction in this interesting science—especially in schools where it is not pursued as a class exercise.

Outlines of Mathematical Science—Davies *1 00

A manual suggesting the best methods of presenting mathematical instruction on the part of the teacher, with that comprehensive view of the whole which is necessary to the intelligent treatment of a part, in science.

Nature & Utility of Mathematics—Davies . .*1 50

An elaborate and lucid exposition of the principles which lie at the foundation of pure mathematics, with a highly ingenious application of their results to the development of the essential idea of the different branches of the science.

Mathematical Dictionary—Davies & Peck .*5 00

This cyclopædia of mathematical science defines with completeness, precision, and accuracy, every technical term, thus constituting a popular treatise on each branch, and a general view of the whole subject.

School Architecture—Barnard *2 25

Attention is here called to the vital connection between a good school-house and a good school, with plans and specifications for securing the former in the most economical and satisfactory manner.

Liberal Education of Women—Orton . . *$1 50

Treats of "the demand and the method;" being a compilation of the best and most advanced thought on this subject, by the leading writers and educators in England and America. Edited by a Professor in Vassar College.

Education Abroad—Northrop *1 50

A thorough discussion of the advantages and disadvantages of sending American children to Europe to be educated; also, Papers on Legal Prevention of Illiteracy, Study and Health, Labor as an Educator, and other kindred subjects. By the Hon. Secretary of Education for Connecticut.

The Teacher and the Parent—Northend . . *1 50

A treatise upon common-school education, designed to lead teachers to view their calling in its true light, and to stimulate them to fidelity.

The Teachers' Assistant—Northend *1 50

A natural continuation of the author's previous work, more directly calculated for daily use in the administration of school discipline and instruction.

School Government—Jewell *1 50

Full of advanced ideas on the subject which its title indicates. The criticisms upon current theories of punishment and schemes of administration have excited general attention and comment.

Grammatical Diagrams—Jewell *1 00

The diagram system of teaching grammar explained, defended, and improved. The curious in literature, the searcher for truth, those interested in new inventions, as well as the disciples of Prof. Clark, who would see their favorite theory fairly treated, all want this book. There are many who would like to be made familiar with this system before risking its use in a class. The opportunity is here afforded.

The Complete Examiner—Stone *1 25

Consists of a series of questions on every English branch of school and academic instruction, with reference to a given page or article of leading text-books where the answer may be found in full. Prepared to aid teachers in securing certificates, pupils in preparing for promotion, and teachers in selecting review questions.

School Amusements—Root *1 50

To assist teachers in making the school interesting, with hints upon the management of the school-room. Rules for military and gymnastic exercises are included. Illustrated by diagrams.

Institute Lectures—Bates *1 50

These lectures, originally delivered before institutes, are based upon various topics in the departments of mental and moral culture. The volume is calculated to prepare the will, awaken the inquiry, and stimulate the thought of the zealous teacher.

Method of Teachers' Institutes—Bates . . . *75

Sets forth the best method of conducting institutes, with a detailed account of the object, organization, plan of instruction, and true theory of education on which such instruction should be based.

History and Progress of Education *1 50

The systems of education prevailing in all nations and ages, the gradual advance to the present time, and the bearing of the past upon the present in this regard, are worthy of the careful investigation of all concerned in education.

THE SCHOOL LIBRARY.

The two elements of instruction and entertainment were never more happily combined than in this collection of standard books. Children and adults alike will here find ample food for the mind, of the sort that is easily *digested*, while not degenerating to the level of modern romance.

LIBRARY OF LITERATURE.

Milton's Paradise Lost. Boyd's Illustrated Ed., $1 60

Young's Night Thoughts . . . . do. . . 1 60

Cowper's Task, Table Talk, &c. . do. . . 1 60

Thomson's Seasons . . . . . . do. . . 1 60

Pollok's Course of Time . . . . do. . . 1 60

These works, models of the best and purest literature, are beautifully illustrated, and notes explain all doubtful meanings.

Lord Bacon's Essays (Boyd's Edition) . . . 1 60

Another grand English classic, affording the highest example of purity in language and style.

The Iliad of Homer. Translated by POPE. . . 80

Those who are unable to read this greatest of ancient writers in the original, should not fail to avail themselves of this metrical version.

Compendium of Eng. Literature—Cleveland, 2 50

English Literature of XIXth Century do. 2 50

Compendium of American Literature do. 2 50

Nearly one hundred and fifty thousand volumes of Prof. CLEVELAND'S inimitable compendiums have been sold. Taken together they present a complete view of literature. To the man who can afford but a few books these will supply the place of an extensive library. From commendations of the very highest authorities the following extracts will give some idea of the enthusiasm with which the works are regarded by scholars:
With the Bible and your volumes one might leave libraries without very painful regret.—The work cannot be found from which in the same limits so much interesting and valuable information may be obtained. — Good taste, fine scholarship, familiar acquaintance with literature, unwearied industry, tact acquired by practice, an interest in the culture of the young, and regard for truth, purity, philanthropy and religion are united in Mr. Cleveland.—A judgment clear and impartial. a taste at once delicate and severe.—The biographies are just and discriminating.—An admirable bird's-eye view.—Acquaints the reader with the characteristic method, tone, and quality of each writer.—Succinct, carefully written, and wonderfully comprehensive in detail, etc., etc.

Milton's Poetical Works—CLEVELAND . . . 2 50

This is the very best edition of the great Poet. It includes a life of the author, notes, dissertations on each poem, a faultless text, and is *the only* edition of Milton with a complete verbal Index.

LIBRARY OF HISTORY.

History of Europe—Alison · · · · · · $2 50

A reliable and standard work, which covers with clear, connected, and complete narrative, the eventful occurrences transpiring from A. D. 1789 to 1815, being mainly a history of the career of Napoleon Bonaparte.

History of England—Berard · · · · · · 1 75

Combining a history of the social life of the English people with that of the civil and military transactions of the realm.

History of Rome—Ricord · · · · · · · · 1 60

Possesses all the charm of an attractive romance. The fables with which this history abounds are introduced in such away as not to deceive the inexperienced reader, while adding vastly to the interest of the work and affording a pleasing index to the genius of the Roman people. Illustrated.

The Republic of America—Willard · · · 2 25

Universal History in Perspective—Willard 2 25

From these two comparatively brief treatises the intelligent mind may obtain a comprehensive knowledge of the history of the world in both hemispheres. Mrs. Willard's reputation as an historian is wide as the land. Illustrated.

Ecclesiastical History—Marsh · · · · · 2 00

A history of the Church in all ages, with a comprehensive review of all forms of religion from the creation of the world. No other source affords, in the same compass, the information here conveyed.

History of the Ancient Hebrews—Mills · · 1 75

The record of "God's people" from the call of Abraham to the destruction of Jerusalem; gathered from sources sacred and profane.

The Mexican War—Mansfield · · · · · · 1 50

A history of its origin, and a detailed account of its victories; with official despatches, the treaty of peace, and valuable tables. Illustrated.

Early History of Michigan—Sheldon · · · 2 50

A work of value and deep interest to the people of the West. Compiled under the supervision of Hon. Lewis Cass. Portraits.

History of Texas—Baker · · · · · · · · 1 25

A pithy and interesting resumé. Copiously illustrated. The State constitution and extracts from the speeches and writings of eminent Texans are appended.

LIBRARY OF BIOGRAPHY.

Life of Dr. Sam. Johnson—Boswell . . .$2 25

This work has been before the public for seventy years, with increasing approbation. Boswell is known as " the prince of biographers."

Henry Clay's Life and Speeches—Mallory
2 vols. 4 50

This great American statesman commands the admiration, and his character and deeds solicit the study of every patriot.

Life & Services of General Scott—Mansfield 1 75

The hero of the Mexican war, who was for many years the most prominent figure in American military circles, should not be forgotten in the whirl of more recent events than those by which he signalized himself. Illustrated.

Garibaldi's Autobiography 1 50

The Italian patriot's record of his own life, translated and edited by his friend and admirer. A thrilling narrative of a romantic career. With portrait.

Lives of the Signers—Dwight 1 50

The memory of the noble men who declared our country free at the peril of their own "lives, fortunes, and sacred honor," should be embalmed in every American's heart.

Life of Sir Joshua Reynolds—Cunningham 1 50

A candid, truthful, and appreciative memoir of the great painter, with a compilation of his discourses. The volume is a text-book for artists, as well as those who would acquire the rudiments of art. With a portrait.

Prison Life 75

Interesting biographies of celebrated prisoners and martyrs, designed especially for the instruction and cultivation of youth.

LIBRARY OF NATURAL SCIENCE.

The Treasury of Knowledge · · · · · ·$1 25

A cyclopædia of ten thousand common things, embracing the widest range of subject-matter. Illustrated.

Ganot's Popular Physics · · · · · · · · 1 75

The elements of natural philosophy for both student and the general reader. The original work is celebrated for the magnificent character of its illustrations, all of which are literally reproduced here.

Principles of Chemistry—Porter · · · · · 2 00

A work which commends itself to the amateur in science by its extreme simplicity, and careful avoidance of unnecessary detail. Illustrated.

Class-Book of Botany—Wood · · · · · · 3 50

Indispensable as a work of reference. Illustrated.

The Laws of Health—Jarvis · · · · · · · 1 65

This is not an abstract *anatomy*, but all its teachings are directed to the best methods of preserving health, as inculcated by an intelligent knowledge of the structure and needs of the human body. Illustrated.

Vegetable & Animal Physiology—Hamilton 1 25

An exhaustive analysis of the conditions of life in all animate nature. Illustrated.

Elements of Zoology—Chambers · · · · · 1 50

A complete view of the animal kingdom as a portion of external nature. Illustrated.

Astronography—Willard · · · · · · · · 1 00

The elements of astronomy in a compact and readable form. Illustrated.

Elements of Geology—Page · · · · · · · 1 25

The subject presented in its two aspects of interesting and important. Illustrated.

Lectures on Natural History—Chadbourne 75

The subject is here considered in its relations to intellect, taste, health, and religion.

LIBRARY OF TRAVEL.

Life in the Sandwich Islands—Cheever . .$1 50

The " heart of the Pacific, as it was and is," shows most vividly the contrast between the depth of degradation and barbarism, and the light and liberty of civilization, so rapidly realized in these islands under the humanizing influence of the Christian religion. Illustrated.

The Republic of Liberia—Stockwell, . . . 1 25

This volume treats of the geography, climate, soil, and productions of this interesting country on the coast of Africa, with a History of its early settlement. Our colored citizens especially, from whom the founders of the new State went forth, should read Mr. Stockwell's account of it. It is so arranged as to be available for a School Reader, and in colored schools is peculiarly appropriate as an instrument of education for the young. Liberia is likely to bear an important part in the future of their race.

Ancient Monasteries of the East—Curzon . 1 50

The exploration of these ancient seats of learning has thrown much light upon the researches of the historian, the philologist, and the theologian, as well as the general student of antiquity. Illustrated.

Discoveries in Babylon & Nineveh—Layard 1 75

Valuable alike for the information imparted with regard to these most interesting ruins, and the pleasant adventures and observations of the author in regions that to most men seem like Fairyland. Illustrated.

A Run Through Europe—Benedict, ... 2 00

A work replete with instruction and interest.

St. Petersburgh—Jermann 1 00

Americans are less familiar with the history and social customs of the Russian people than those of any other modern civilized nation. Opportunities such as this book affords are not, therefore, to be neglected.

The Polar Regions—Osborn 1 25

A thrilling and intensely interesting narrative of one of the famous expeditions in search of Sir John Franklin—unsuccessful in its main object, but adding many facts to the repertoire of science.

Thirteen Months in the Confederate Army 75

The author, a northern man conscripted into the Confederate service, and rising from the ranks by soldierly conduct to positions of responsibility, had remarkable opportunities for the acquisition of facts respecting the conduct of the Southern armies, and the policy and deeds of their leaders. He participated in many engagements, and his book is one of the most exciting narratives of adventure ever published. Mr. Stevenson takes no ground as a partizan, but views the whole subject as with the eye of a neutral—only interested in subserving the ends of history by the contribution of impartial facts. Illustrated.

56

LIBRARY OF REFERENCE.

Home Cyclopædia of Literature & Fine Arts $3 00
A complete index to all terms employed in belles lettres, philosophy, theology, law, mythology, painting, music, sculpture, architecture, and all kindred arts.

The Rhyming Dictionary—Walker 1 25
A serviceable manual to composers, being a complete index of allowable rhymes.

The Topical Lexicon—Williams 1 75
The useful terms of the English language *classified by subjects* and arranged according to their affinities of meaning, with etymologies, definitions and illustrations. A very entertaining and instructive work.

Mathematical Dictionary—Davies & Peck . 5 00
A thorough compendium of the science, with illustrations and definitions.

RELIGIOUS LIBRARY.

The Service of Song—Stacy $1 50
A treatise on Singing, in public and private devotion. Its history, office, and importance considered.

True Success in Life—Palmer $1 50
Earnest words for the young who are just about to meet the responsibilities and temptations of mature life.

"Remember Me"—Palmer 1 50
Preparation for the Holy Communion.

Chrysostom, or the Mouth of Gold—Johnson 1 00
An entertaining dramatic sketch, by Rev. Edwin Johnson, illustrating the life and times of St. Chrysostom.

The Memorial Pulpit—Robinson. 2 vols., each 1 50
A series of wide-awake sermons by the popular pastor of the Memorial Presbyterian Church, New York.

Responsive Worship—Budington 60
An argument in favor of alternate Scripture reading by Pastor and Congregation.

Lady Willoughby 1 00
The diary of a wife and mother. An historical romance of the seventeenth century. At once beautiful and pathetic, entertaining and instructive.

Favorite Hymns Restored—Gage 1 25
Most of the standard hymns have undergone modification or abridgment by compilers, but this volume contains them exactly as written by the authors.

Poets' Gift of Consolation 1 50
A beautiful selection of poems referring to the death of children.

VALUABLE LIBRARY BOOKS.

The Political Manual—Mansfield $1 25

Every American youth should be familiar with the principles of the government under which he lives, especially as the policy of this country will one day call upon him to participate in it, at least to the extent of his ballot.

American Institutions—De Tocqueville . . 1 50
Democracy in America—De Tocqueville . . 2 50

The views of this distinguished foreigner on the genius of our political institutions are of unquestionable value, as proceeding from a standpoint whence we seldom have an opportunity to hear.

Constitutions of the United States 2 25

Contains the Constitution of the General Government, and of the several State Governments, the Declaration of Independence, and other important documents relating to American history. Indispensable as a work of reference.

Public Economy of the United States . . . 2 25

A full discussion of the relations of the United States with other nations, especially the feasibility of a free-trade policy.

Grecian and Roman Mythology—Dwight . 3 00

The presentation, in a systematic form, of the Fables of Antiquity, affords most entertaining reading, and is valuable to all as an index to the mythological allusions so frequent in literature, as well as to students of the classics who would peruse intelligently the classical authors. Illustrated.

General View of the Fine Arts—Huntington 1 75

The preparation of this work was suggested by the interested inquiries of a group of young people concerning the productions and styles of the great masters of art, whose names only were familiar. This statement is sufficient index of its character.

The Poets of Connecticut—Everest 1 75

With the biographical sketches, this volume forms a complete history of the poetical literature of the State.

The Son of a Genius—Hofland 75

A juvenile classic which never wears out, and finds many interested readers in every generation of youth.

Sunny Hours of Childhood 75

Interesting and moral stories for children.

Morals for the Young—Willard 75

A series of moral stories, by one of the most experienced of American educators. Illustrated.

Improvement of the Mind—Isaac Watts . . 50

A classical standard. No young person should grow up without having perused it

PUBLIC WORSHIP.

Songs for the Sanctuary, ... $2 50

By REV. C. S. ROBINSON. 1344 Hymns, with Tunes. The most successful modern hymn and tune-book, for congregational singing. More than 200,000 copies have been sold. Separate editions for Presbyterian, Congregational, and Baptist Churches. Editions without Tunes, $1.75; in large type, $2.50. Abridged edition ("Songs for Christian Worship"), 859 Hymns, with Tunes, $1.50. Chapel edition, 607 Hymns, with Tunes, $1.40.

International Singing Annual, ... 25

Metrical Tune Book, ... 1 00

To be used with any hymn-book. By PHILIP PHILLIPS.

Baptist Praise Book, ... 2 50

By REV. DRS. FULLER, LEVY, PHELPS, FISH, ARMITAGE, WINKLER, EVARTS, LORIMER and MANLY, and J. P. HOLBROOK, Esq. 1311 Hymns, with Tunes. Edition without Tunes, $1.75. Chapel edition, 550 Hymns, with Tunes, $1.25.

Plymouth Collection, ... 2 50

(Congregational.) By REV. HENRY WARD BEECHER. 1374 Hymns, with Tunes. Separate edition for Baptist Churches. Editions without Tunes, $1.25 and $1.75.

Hymns of the Church, ... 2 75

(Undenominational.) By REV. DRS. THOMPSON, VERMILYE, and EDDY. 1007 Hymns, with Tunes. The use of this book is required in all congregations of the Reformed Church in America. Edition without Tunes, $1.75. Chapel edition ("Hymns of Prayer and Praise"), 320 Hymns, with Tunes, 75 cts.

Episcopal Common Praise, ... 2 75

The Service set to appropriate Music, with Tunes for all the Hymns in the Book of Common Prayer.

Hymnal, with Tunes, ... 1 25

(Episcopal.) By HALL & WHITELEY. The new Hymnal, set to Music. Edition with Chants, $1.50. Edition of Hymns only ("Companion" Hymnal), 60 cts.

Quartet and Chorus Choir, ... 3 00

By J. P. HOLBROOK. Containing Music for the Unadapted Hymns in Songs for the Sanctuary.

Christian Melodies. By GEO. B. CHEEVER. Hymns and Tunes. 1 00

Mount Zion Collection. By T. E. PERKINS. For the Choir. 1 25

Selah. By THOS. HASTINGS. For the Choir. 1 25

Public Worship (Partly Responsive) $1 00

Containing complete services (not Episcopal) for five Sabbaths; for use in schools, public institutions, summer resorts, churches without a settled pastor; in short, wherever Christians desire to worship—no clergyman being present.

The Union Prayer Book, ... 2 50

A Manual for Public and Private Worship. With those features which are objectionable to other denominations of Christians than Episcopal eliminated or modified. Contains a Service for Sunday Schools and Family Prayers.

The Psalter, ... 16mo, 60 cts.; 8vo, 90

Selections from the Psalms, for responsive reading.

FURNITURE.

(SUPPLIED BY THE NATIONAL SCHOOL FURNITURE CO.)

PEARD'S PATENT FOLDING DESK AND SETTEE.

This great improvement for the school-room has come already into such astonishing demand as to tax the utmost resources of the company's two factories to supply it. By a simple movement the desk-lid is folded away over the back of the settee attached in front, making a false back, and at once converting the school-room into a lecture or assembly-room. When the seat also is folded, the whole occupies *only ten inches of space*, leaving room for gymnastic exercises, marching, etc., or for the janitor to clean the room effectively.

NATIONAL STUDY DESK AND SETTEE.

When not in use for writing, the desk-lid slides back vertically into a chamber, leaving in front an "easel," with clamps, upon which the student places his book and studies in an erect posture. As a folding-desk this offers many of the same advantages as the "Peard."

THE GEM DESK AND SETTEE.

Fixed top, and folding seat. This is the *neatest* pattern of the Standard School Desk, and the *strongest* in use.

THE ECONOMIC DESK AND SETTEE.

This is the *cheapest* good desk, with stationary lid and folding seat.

All descriptions of

HIGH SCHOOL DESKS,   SCHOOL SETTEES,
TEACHERS' DESKS,   CHURCH SETTEES,
BLACKBOARDS,   PEW ENDS,
CHAIRS,   LECTERNS, Etc.

Also,

TAYLOR'S PATENT

CLASS AND LECTURE CHAIR.

The difficulty of reconciling furniture appropriate for the Lecture-room or Church with that convenient for the Sunday-school is an old one. This article effectually remedies it. It consists simply of a plan by which chairs of a somewhat peculiar shape are connected with a coupling. The rows of chairs thus adjusted may at pleasure and with ease be spread out straight in one line, forming pews or benches; or they may be bent in an instant into a semi-circular form to accomodate classes of any size to receive instruction from teachers seated in their midst.

For further particulars, consult catalogues of the National School Furniture Co. and the Taylor Patent Chair Co., which may be obtained of A. S. Barnes & Co.

The Peabody Correspondence.

NEW YORK, April 29, 1867.

To THE BOARD OF TRUSTEES OF THE PEABODY EDUCATIONAL FUND:

GENTLEMEN—Having been for many years intimately connected with the educational interests of the South, we are desirous of expressing our appreciation of the noble charity which you represent. The Peabody Fund, to encourage and aid common schools in these war-desolated States, cannot fail of accomplishing a great and good work, the beneficent results of which, as they will be exhibited in the future, not only of the stricken population of the South, but of the nation at large, seem almost incalculable.

It is probable that the use of meritorious text-books will prove a most effective agency toward the thorough accomplishment of Mr. Peabody's benevolent design. As we publish many which are considered such, we have selected from our list some of the most valuable, and ask the privilege of placing them in your hands for gratuitous distribution in connection with the fund of which you have charge, among the teachers and in the schools of the destitute South.

Observing that the training of teachers (through the agency of Normal Schools and otherwise) is to be a prominent feature of your undertaking, we offer you for this purpose 5,000 volumes of the "Teachers' Library,"—a series of professional works designed for the efficient self-education of those who are in their turn to teach others—as follows:—

500 Page's Theory and Practice of Teaching.	250 Bates' Method of Teachers' Institutes
500 Welch's Manual of Object-Lessons.	250 De Tocqueville's American Instit'ns
500 Davies' Outlines of Mathematical Science.	250 Dwight's Higher Christian Educat'n.
	250 History of Education.
250 Holbrook's Normal Methods of Teaching.	250 Mansfield on American Education.
	250 Mayhew on Universal Education.
250 Wells on Graded Schools.	250 Northend's Teachers' Assistant.
250 Jewell on School Government.	250 Northend's Teacher and Parent.
250 Fowle's Teachers' Institute.	250 Root on School Amusements.
	250 Stone's Teachers' Examiner.

In addition to these we also ask that you will accept 25,000 volumes of school-books for intermediate classes, embracing—

5,000 The National Second Reader.	5,000 Beers' Penmanship.
5,000 Davies' Written Arithmetic.	500 First Book of Science.
5,000 Monteith's Second Book in Geography.	500 Jarvis' Physiology and Health.
	500 Peck's Ganot's Natural Philosophy.
3,000 Monteith's United States History.	500 Smith & Martin's Book-keeping.

Should your Board consent to undertake the distribution of these volumes, we shall hold ourselves in readiness to pack and ship the same in such quantities and to such points as you may designate.

We further propose that, should you find it advisable to use a greater quantity of our publications in the prosecution of your plans, we will donate, for the benefit of this cause, *twenty-five per cent.* of the usual wholesale price of the books needed.

Hoping that our request will meet with your approval, and that we may have the pleasure of contributing in this way to wants with which we deeply sympathize, we are, gentlemen, very respectfully yours,   A. S. BARNES & CO.

BOSTON, *May* 7, 1867.

MESSRS. A. S. BARNES & Co., PUBLISHERS, NEW YORK:

GENTLEMEN—Your communication of the 29th ult., addressed to the Trustees of the Peabody Education Fund, has been handed to me by our general agent, the Rev. Dr. Sears. I shall take the greatest pleasure in laying it before the board at their earliest meeting. I am unwilling, however, to postpone its acknowledgment so long, and hasten to assure you of the high value which I place upon your gift. Five thousand volumes of your "Teachers' Library," and twenty-five thousand volumes of "School-books for intermediate classes," make up a most munificent contribution to the cause of Southern education in which we are engaged. Dr. Sears is well acquainted with the books you have so generously offered us, and unites with me in the highest appreciation of the gift. You will be glad to know, too, that your letter reached us in season to be communicated to Mr. Peabody, before he embarked for England on the 1st inst., and that he expressed the greatest gratification and gratitude on hearing what you had offered.

Believe me, gentlemen, with the highest respect and regard, your obliged and obedient servant,   ROBT. C. WINTHROP, Chairman.

GENERAL INDEX TO
A. S. Barnes & Co.'s Descriptive Catalogue,

www.ingramcontent.com/pod-product-compliance
Lightning Source LLC
Chambersburg PA
CBHW021819190326
41518CB00007B/660